CONSCIOUSNESS AND TIME

A NEW APPROACH

RAFAEL PINTOS-LÓPEZ

For Inés, who had to put up with the lunacy and the obsession.

CONTENTS

Introduction	xi
CHAPTER 1	1
Consciousness	1
CHAPTER 2	22
Time	22
CHAPTER 3	43
Objective reality & quantum physics (an optional chapter)	43
CHAPTER 4	61
Greek philosophers and the Bible (another optional chapter)	61
CHAPTER 5	69
St. Augustine	69
CHAPTER 6	78
Newton and Leibniz	78
CHAPTER 7	84
Jorge Luis Borges	84
CHAPTER 8	89
Einstein and Schrödinger	89
CHAPTER 9	97
Stephen Hawking and Carlo Rovelli	97
CHAPTER 10	105
Conclusion	105
Afterword	117
Acknowledgments	137

"I've never believed that they're separate. Leonardo da Vinci was a great artist and a great scientist. Michelangelo knew a tremendous amount about how to cut stone at the quarry. The finest dozen computer scientists I know are all musicians. Some are better than others, but they all consider that an important part of their life. I don't believe that the best people in any of these fields see themselves as one branch of a forked tree. I just don't see that. People bring these things together a lot. Dr Land at Polaroid said, 'I want Polaroid to stand at the intersection of art and science', and I've never forgotten that. I think that that's possible, and I think a lot of people have tried."

- Steve Jobs, interview, *Time magazine*, 10 October 1999.

"I seem to feel Napoleon's influence on our quiet evening in the garden for instance — I think I see for a moment how our minds are all threaded together — how any live mind today is of the very same stuff as Plato's & Euripides. It is only a continuation & development of the same thing. It is this common mind that binds the whole world together; & all the world is mind."

- Virginia Woolf, *Journal*, 1 July 1903.

"Thinking outside the brain means skilfully engaging entities external to our heads — the feelings and movements of our bodies, the physical spaces in which we learn and work, and the minds of the other people around us — drawing them into our own mental processes. By reaching beyond the brain to recruit these "extra-neural" resources, we are able to focus more intently, comprehend more deeply, and create more imaginatively — to entertain ideas that would be literally unthinkable by the brain alone."

- Anne Murphy Paul, *The extended mind: the power of thinking outside the brain*, 2022.

"I am not sure that I exist, actually. I am all the writers that I have read, all the people that I have met, all the women that I have loved; all the cities that I have visited, all my ancestors."

- Jorge Luis Borges, interview, *El País*, 1981

EPIGRAPH

The title of this book is "Consciousness and Time", not the other way around, "Time and Consciousness". There is a reason for that. It follows the Augustinian principle that "time is an extension of the soul". Time is subordinate to consciousness. Without consciousness there is no time.

The subtitle of this book includes the words "a new approach". I do not claim that this work provides a breakthrough of any kind. It suggests options that have not been considered.

The book does not analyse consciousness. It proposes a taxonomic approach to its study, a two-tiered system of human consciousness. That taxonomy has several consequences, *inter alia*, a new approach to the essence of time.

I can only hope the ideas are useful, that's all.

As the title suggests, the book provides a new approach to the study of something as important as the functioning of thought and human life. It attempts to refresh insights of philosophers as universally important as Aristotle and St Augustine. The book questions current scientific methodology and introduces new ideas on the subject of consciousness and time.

R P-L

INTRODUCTION

> *"We no longer live in Newton's mechanical universe -*
> *we live in a banana peel universe, and never*
> *will we be able to know anything,*
> *control anything or predict anything."*
> *- Emily Levine*

This is not a book about philosophy, or science, or history. It's a bit of everything. I propose some ideas. More than anything, the book deals with two closely related topics that should be of interest to everyone. It also deals with different thoughts and perspectives on consciousness and time. Those two concepts have avoided a definition and seem to mix with each other. They appear to be outside the orbits of science and philosophy, or perhaps they are part of both.

My training at the tertiary level is in theoretical linguistics, but I have worked mostly within applied linguistics. I am a translator and I have been a translator for more than half a century, which means that I am not accredited to give scientific or philosophical opinions. However, like many others, I have a profound curiosity about consciousness and time.

If you learn something by reading this (as much as I learned while I wrote it), I will have achieved my goal.

Although theoretical linguistics is sometimes considered a science, translation is, more than anything, an art. There is always magic. One always finds hidden meanings. When you move from one language to another there is almost never a total coincidence between the terms. What really manages to transfer the meaning is, more than anything, something similar to artistic creation. Applied linguistics, then, is in that overlapping space situated right between science and art.

I admire what science has done and does, although many of the discoveries of science seem esoteric, sometimes even miraculous. In their apparent lack of common sense, those discoveries increasingly resemble religion.

Example: on the event horizon of a black hole the particles are two-dimensional. An idea that appears like dogma that you reach by doing magic with numbers and letters. One can believe or not believe what scientists say, but we can be sure of one thing: we will never experience or know for sure what happens on the edge of a black hole. What is a black hole? What is an event horizon? What are two-dimensional particles? We wonder.

There are many black holes in the universe. They are mass concentrations with such gravitation that it attracts everything that happens near them. What enters a black hole disappears.

In the story *The Disk*, Jorge Luis Borges imagines a two-dimensional object, Odin's disk:

"... He looked me in the eye.

"You may touch it." I had my doubts, but I reached out and with my fingertips I touched his palm. I felt something cold, and I saw a quick gleam. His hand snapped shut. I said nothing.

"It is the disk of Odin," the old man said in a patient voice, as though he were speaking to a child. "It has but one side. There is not another thing on earth that has but one side."

Möbius' tape, which has only one side, is an example, but compared to the two-dimensional particles of the event horizon, it's rather like a childish trick performed by a cheap magician. The difference is that a magic trick can be explained, the two-dimensionality of a particle is something very difficult to imagine or understand.

Scientists give us a simplified example of an event horizon. You have to imagine the top page of a sheet of paper. On the sheet we put a cylinder—a three-dimensional world. The perimeter of the circle that touches the sheet of paper is the equivalent to the event horizon of the black hole. According to the formulas, there—we know for sure—the particles are two-dimensional.

According to scientists, we only know the mass, electric charge, and angular momentum of the black hole, but we cannot know the structure and nature of the material that has formed the black hole. We only have uncategorised quantitative information.

All that sounds like a mystery to me: the Holy Trinity. Of course, it can be demonstrated with formulas; and, according to some physicists, the nature of time has to do with the difference between the particles that are on the surface of a black hole and those that are inside. Hopefully, before I finish this book, I will be able to explain something about time, and it's not going to be like that at all.

All that, you will see, are ideas that occurred to a physicist, or a group of physicists, and that he—or they—proposed by means of formulas. In reality, although it is feasible that at some point it may have some usefulness, everything related to the information (?) that can be lost inside a black hole is something totally meaningless to the rest of us mortals.

It seems that, from the very beginning, I'm questioning what physicists say. No, it's not like that at all. I will try to show that everything should be falsifiable.

Incidentally, what happens is that when we learn something about science, or philosophy, or semantics, they teach us to question and ask ourselves why something happens and how it happens. Sure. The interesting thing, too, is the degree of relevance of any phenomenon that is studied. On top of that, we appear to be limited by our senses, which are very limited.

Sabine Hossenfelder—a specialist in quantum mechanics—tells us, on the basis of a prediction by Laplace: *"The laws of nature organise matter in the universe in different configurations, but from the configuration at a given time you can find out what the configuration was at an earlier time, at least in theory. In practice, it usually can't be done because you don't have all the details. So, if someone dies, all the information about them is dispersed in a way with which there is no longer possible communication. But, in principle, the data cannot be destroyed—they are present in the correlations between the atoms and the light quantum and are slowly dispersed through the solar system and the entire universe."*

This makes me think of two possibilities. For example:

1) We know what Bernard Shaw thought about certain topics because he left his thoughts in writing. So, in the future, we could possibly rebuild the neurons, dendrites, myelin, etc., as they were in his brain at a particular time. Does that mean that we could know what he thought about that topic—if he did—when he was twenty years old? The proposition seems to me at least questionable: by observing Bernard Shaw's reconstructed brain we would have changed those particles, just by observing them. On the one hand, quantum mechanics tells me that a configuration can be reconstructed to some extent; on the other, the Uncertainty Principle applies.

2) Photography is based on the moment when light touches an object. A hologram reproduces the moment when light is emitted towards the object. Theoretically, if we travelled in space two hundred and seven light years and managed to capture a ray of light over Belgium, could we build a hologram of the Battle of Waterloo, which occurred in 1815? The proposition is also very questionable, because there was no observer from above, which means that an aerial image of the battle does not exist. Without an observer there is no phenomenon.

There are countless possibilities for reconstructing configurations that can never be carried out for one reason or another. Unless it is another type of information that I do not know.

Quantum mechanics tells me that I can't get information about the particles that are inside a black hole. How is it possible to know that with any degree of certainty, if physical verification is impossible? Scientific conjectures of that type become religious articles of faith.

According to Hossenfelder—and according to Einstein's space-time theories, that is, the so-called "block universe" or "eternalism"—past and present exist. Future, doesn't. Again, I beg to disagree, to some extent. There is only present.

What is illogical about all this is not that my present can coexist with another person's different present, if the person is on the other side of the world—for example—since that is verifiable. We can say, according to the clock, that the person is in another time [zone], but I can talk to her, and see her in my present (and in her present). What doesn't make sense is for anyone to tell me that the past exists just like the present—according to the Theory of Relativity—but that the future doesn't, because of the improbability posed by quantum mechanics. Perhaps we could say that the future does not exist as yet.

Hossenfelder knows very well how to express her doubts about current affairs in physics: *"Imagine you go to a zoology conference. The first speaker talks about her 3D model of a 12-legged purple spider that lives in the Arctic. There's no evidence it exists, she admits, but it's a testable hypothesis, and she argues that a mission should be sent off to search the Arctic for spiders.*

The second speaker has a model for a flying earthworm, but it flies only in caves. There's no evidence for that either, but he petitions to search the world's caves. The third one has a model for octopuses on Mars. It's testable, he stresses.

Kudos to zoologists, I've never heard of such a conference. But almost every particle physics conference has sessions just like this, except they do it with more maths. It has become common among physicists to invent new particles for which there is no evidence, publish papers about them, write more papers about these particles' properties, and demand the hypothesis be experimentally tested. Many of these tests have actually been done, and more are being commissioned as we speak. It is wasting time and money".
—Hossenfelder maintains. I totally agree.

Karl Popper said that, for any scientific idea to be good, it had to be able to be falsifiable. According to Hossenfelder, some physicists now seem to have understood that any idea that can be falsifiable is good science. That's not the case, the prediction of a new particle is only necessary when it solves a problem. It's not that all scientists are doing it, but some do. Hossenfelder appears to be the girl who points out that the emperor is naked. A new paradigm is needed.

I was going to focus on consciousness and time and have gone on to talk about the way science wastes resources studying certain phenomena; how it digresses. Am I digressing too?

Actually, no. Here I wish to talk about consciousness and time, but also about the impossibility to advance in the study of those topics with the methods we are using. We continue to try to solve certain problems the same way, although it appears evident that issues such as consciousness and time constitute something akin to the limit of science as we understand it in the West.

At the conclusion of this book, I'm going to propose several new ideas that, I hope will help in the development of research on consciousness and time and that, as far as I know, have not been considered by science or philosophy. The two most important ones, which are the main reason for this book, are:

1) that we have two types of consciousness; but that they are not a continuum. They are discrete. What this means is that research on one of them does not necessarily help when we wish to find something about the nature of the other.

2) as a result of the first idea, I propose a new approach to the way we consider time. I propose that time is only a human device; not that it does not exist. Only that it is a human device that exists within certain limitations.

This book is not an attempt to reinvent the wheel. Both Aristotle and St Augustine linked consciousness and time. It's just that they did not have the tools to reach the conclusions we can reach nowadays. Their reasoning followed different paths.

In 1905, Albert Einstein, a young employee of the Geneva Patent Office and a fan of physics, proposed the Theory of Relativity, which questioned Newtonian mechanics and revolutionised Western thought. Gravity exists and bodies are attracted, but the movements of one and the other are relative. Without going into details, one of the ideas that relativity establishes—although not clearly—is that objective reality is questionable.

In addition, Einstein's brilliant theory is based on an absolutely counter-intuitive innovation. Space is perceived with the senses, while time is perceived only through high consciousness. Einstein unites the two concepts and begins to talk about space-time.

Scientists understand and applaud Einstein's theory, but their acceptance does not go as far as saying that objective reality is questionable. The observations, experiments and achievements of Western science continue to be based on the notion of an objective and immutable reality.

In 1492, Christopher Columbus said that you could reach the Indies by going West. Some thought the idea was crazy. The Earth is flat, something totally obvious. If you go West, you reach the end of the ocean sea and ships fall into the void. A few years later, against all logic—and going towards the unknown at the risk of their lives—Ferdinand Magellan, who died in the attempt, and Juan Sebastián el Cano, who succeeded, managed to circumnavigate the world.

In the 17th century, Galileo Galilei, a Florentine who followed Copernicus, another rebel—moved away from geocentrism and declared that the Earth revolved around the Sun—establishing that humanity is not the centre of the universe. He did that intuitively. The reality at the time was that the Sun obviously revolved around our planet. Galileo had a hunch: "What if..."?

Brunelleschi is a goldsmith who, against all ideas available at the time builds the Duomo of Santa María del Fiore, in Florence, using a method that the connoisseurs of the time thought totally impossible. The Duomo was built without the help of internal wooden beams, circle by circle, going up and approaching the centre until it reached the top. The bricks were placed using a fishbone pattern and builders had to wait for each row to dry before laying the next one. The dome consists of two layers, one inside and one outside. The system is totally intuitive and, as we already said, it opposed the engineering ideas of the time. The Duomo, still erect and solid in the centre of Florence, is a monument to the vision of someone who thought differently.

As I write this, I recall Heidegger, who believed that "science cannot think." The intention of science is to think about reality, and do so by going directly to the essence of things. That way, it surpasses the possibilities of thinking.

Perhaps our chances of pondering consciousness and time are limited at least—as we shall see—from the perspective and within the methods of current Western science. In that sense, Wittgenstein established a principle that can be applied to both philosophy and science: "When you cannot talk about something, it is better to remain silent." What Wittgenstein was suggesting to us was that there are things that are ineffable. Are consciousness and time ineffable? We don't know yet. In the meantime, let's consider possibilities using common sense.

Naturally, the mind discards everything unnecessary and focuses on a single thing, on a single concept, to be able to think about it. Science—and especially its quintessence, mathematics—abstracts to be able to analyse and contemplate the nuclear component of a topic. In doing so, it impoverishes and removes dynamics and reality from concepts. Heidegger contrasts experience, which is rich, with the poverty of the scientific description of it. It would seem that there are things that are impossible to abstract, especially when you try to understand them better. Borges says that synthesis is a way of killing.

Among Western philosophers, Leibniz, for example, specifies that unity is a prerequisite for something to be considered a being. Something exists when it exists within certain limits, and when everything around it is not that something. By establishing that limitation, and applying the principles of Western philosophy, Leibniz unwittingly denies the existence of consciousness (unique and multiple at the same time), and of time, which we can measure, but whose limits and essence are unknown. That would also include the human being, who has limits (the skin), but whose consciousness does not have them.

Heidegger says that a being, according to experience, has to be multiple, otherwise it is not a being. We know that mathematics multiplies entities. Unity would seem to be, for us at least, an illusory effect of multiplicity.

We can only access the collective consciousness through our individual consciousness. But Western thought does not seem to allow us to identify unity with multiplicity. Mathematics contradicts that possibility.

Aristotle understood that time was the measure of change, the measure of how things kept on transforming themselves (we will check that out). By the 17th century, philosophers believed that time was absolute, that it had nothing to do with change, that it was independent of the rest. Time was part of Newton's objective reality and in those days people swore it was that way. Since then, we learned a lot. At one point, Einstein came out and said that time—and in turn space—was relative, that there was no absolute time. (We will also see about those other two possibilities).

What does Einstein mean that time is relative? Well, depending on where the observer is, what he is doing, or if he is going from one place to another, time changes.

I've digressed; that happens to me often, you'll soon find out. But let's go back to our topic. In the West, we know—or we think we know—that time moves forward (unlike what the Incas thought, who believed that time moves backwards). They said that we look forward, that's why we know the past. What we can't see—the future—is behind our backs, and we can't see it because we don't have eyes on the back of our neck, they thought.

Time, however, is not perceived through the senses, but through high consciousness. The senses help us perceive matter, the physical reality of which we are part. Consciousness, something that also has an apparent physical origin, is not part of the material world. Time, also intangible, is intimately related to it, as we shall see. St Augustine, a great thinker, also realised there was a link between consciousness and time.

CHAPTER 1

CONSCIOUSNESS

> *"You can't write directly about the soul.*
> *When you look at it, it fades."*
> - Virginia Woolf
>
> *"We are the way the cosmos knows itself."*
> - Carl Sagan

I start this chapter quoting two thinkers from different fields who go far beyond poetry and far beyond science. Somehow, they have a similar approach to the concept. Consciousness is something much greater than we normally imagine. Virginia Woolf tells us that the soul vanishes when we look at it. A bit of quantum physics in that the observer and the observed are mutually influenced, but also the acceptance that consciousness is something much greater than a subject and an object of observation. Carl Sagan goes even further: we're part of a cosmic consciousness, or we are *the* cosmic consciousness, he tells us.

And maybe it's like that. We are part of the cosmos. Our consciousness, individual or collective, so to speak, is part of the cosmos. Perhaps the only cosmic consciousness? We don't know.

Later we will see why consciousness is really the last frontier of science.

The same happens with Schrödinger, who thought that every human being is the same as any other human being who has lived before them. Looking at the mountains, he imagined other men, who had seen those mountains thousands of years ago. He questioned: why think that one's experience is different from theirs?

Jorge Luis Borges used to say that killing one man is like killing all men, adding that the suffering of one person is like the suffering of the whole world. It cannot be multiplied by the number of people. One suffering is all the suffering there is; the intensity is the same.

Human beings have an incredible capacity to share emotion. Of course, we can share much more than emotion.

On 18 December 2022, Gonzalo Montiel, turned around and took his jersey off, running towards his team mates. The Argentine player had kicked the last penalty of the 2022 Soccer World Cup, twelve yards from the French goal, and the ball had gone beyond the goal line. The Argentine team were champions of the world.

That very instant, millions of people went out to celebrate, many of them in *albiceleste* jerseys, jumping for joy, singing, dancing, and cheering the names of the Argentine players. In Buenos Aires alone, about five million celebrated on Avenida 9 de Julio and nearby streets. Many million went out in Dacca, flying Argentine flags, and in the whole of Bangladesh they were probably many more, as in various other parts of the world. At that point, some of them had probably dreamed and forgotten their individual dreams. All of them saw different things and experimented different realities and different points of view, from emergency settlements to sumptuous neighbourhoods in Argentina, or in European or American cities. The celebration lasted several days. The emotion of what had happened was shared by all of them.

There is no doubt that two or more people can experience almost identical feelings. Fear, love, hatred, any sensation caused by a tribal or family confrontation; all of them are shared with different levels of intensity, but they are shared. The consciousness where those feelings are appears to be individual, but the emotion goes beyond the skin of the person who feels it; that emotion is as abstract as consciousness itself. What is extremely interesting is that our species has evolved to share much more than emotions. We communicate, we share knowledge among billions of people. News are shared instantaneously around the world. Our common consciousness is enormous. How did it grow? Well, it is complicated, but it can be understood. We are beginning to make inroads into the mystery.

According to science, if we want to define something we have to explain where it came from, how it appeared and, to reach a satisfactory conclusion, we have to assume the existence of things, because we can never be sure of the origins of those facts or those things. In short, that certain things that seem real to us may not be so real.

For example, how do we have a consciousness? Where did it come from? Did it evolve? Are we the only living beings who have a consciousness? Is our consciousness similar to that of other animals? Obviously, our senses are different. A dog smells much better than us, and a bird can see much better. But our consciousness has evolved. We know that we live and we know that we die. We can communicate very complex ideas.

This brings us to the problem that we do not know the origin of consciousness, nor of time for that matter, nor do we know how they operate. We don't know what consciousness is. We can't define it. Nowadays there are scientists who say that time is not real, and that consciousness is not real either, that they do not exist. Well, perhaps it's not quite like that. Maybe we have to analyse why they say it. Humbly, I think we would have to start by assuming that, although intangible, time exists at some level, at least for us, at our everyday level. Consciousness, as well—intangible as it is—is also real, no doubt.

If it weren't, we wouldn't be looking at these pages or understanding what these words mean. As I write this, I realise that science does one thing, and one thing only: the study of the universe. It does not seek meaning. It does not seek a purpose. It wants to ignore consciousness. At least thus far it has tried to do so. Until the moment when human beings, as a species, acquired reason, the universe lacked meaning. Now it has meaning, or it appears to be acquiring it.

NEUROLOGIST PATRICK HOUSE begins a chapter of his book, *Nineteen ways of looking at consciousness*, with a series of questions: *"Where does consciousness end and the rest of the world begin? Where is the line between inside and outside? Between life and not life? Between the parts of the universe that are conscious and those that are not? Between you and not you?"* The questions are very complicated, but maybe they are not the questions to ask. Talking about *"an individual consciousness"* may be a misleading way to begin studying the subject. Perhaps the answer is the quote I make from Sagan at the beginning of this chapter: *"We are the way the cosmos knows itself."*

Above, I discuss quantum physics and consciousness, the meaning that science does not seek. Perhaps my disenchantment with science has to do with the way it is approaching the pursuit of knowledge. At the moment, for example, physics is totally concentrated on the search for new particles. I believe that what is necessary is a paradigm shift: by studying consciousness, science may reach what is hitherto unknowable. Something that, surely, my generation will not see. By observing something, the observer changes the object of his attention, we said. Apparently, that happens to particles at a very small level. But consciousness cannot be observed either. And when you can glimpse a little piece of consciousness, consciousness changes. St. Augustine says so, Virginia Woolf says so, Schrödinger says so, quantum mechanics says so. Without an observer, there is no object. But there is a relationship. The object and the observer are part of the same reality. They are connected. They are as entangled as photons at the quantum level. Lately there has been a lot of talk about the subject, which is very complicated. Not even a definition of consciousness has yet been reached.

We know that we think; we know that we understand many things, that we perceive sights, sounds, smells, and that we can translate that, for example, into emotions, but we do not know very well how it happens. Descartes proposed to us the principle of rationalism by saying, "I think, therefore I am." You can talk about the accuracy of the translation of that sentence into English, but that's beside the point. What comes to mind is that it made us think about consciousness and existence. And about the existence of consciousness and its relationship with being. We'll talk about that later. When Charles Darwin proposed the theory of evolution, there was another great scientist, Alfred Wallace, who simultaneously came to the same conclusion. Many consider him a co-author of evolutionism. The difference is that Wallace had doubts about natural selection. He thought that natural selection was insufficient in itself to explain a number of human characteristics, including consciousness.

Wallace did not agree all his life with the theory of natural selection. When he reached old age, he began to return to his religious notions. What struck him most about Darwin's idea was the way the brain—but especially the human mind, which is not the same—could have evolved according to that explanation. By then, Wallace had already renounced materialistic modes of explaining consciousness. It was impossible for him to explain the complexity of the human mind with the limited means of natural selection.

Many scientists are still looking for consciousness in the brain. They are looking for it within the physical, within matter. In that sense, science sins for lack of imagination. I give you a very real example of lack of imagination: In 1990, Derek Bickerton, in his book *Language and species*, says, *"It is hard to imagine how a creature without language would think, but one may suspect that a world without any kind of language would in some ways resemble a world without money —a world in which actual commodities, rather than metal or paper symbols for the value of these, would have to be exchanged. How slow and cumbersome the simplest sale would be, and how impossible the more complex ones!"*

One can agree with the improbability of thinking without language, although we now know that language is only part of thought, just as we know that the world can function without physical money. So much so that, today—thirty-two years after Bickerton's book was published—I manage very well without *"metal or paper symbols"* in my pocket. I don't have plastic *"symbols"* either. I pay for everything I buy with my phone, or my watch, which are wirelessly connected to bank accounts. And I get paid for my services the same way. I owe these advances to a man who had a special kind of imagination. His name was Steve Jobs, and he said his ideal place was the intersection between design and technology.

Human beings continue to search for a reason for existence. Science—the creation of the human mind—tells us, on the other hand, that there is no reason for existence. We have no meaning. We only exist. Indeed, evolutionists find that explaining consciousness and its desire to find meaning is something rather difficult. It's a problem that, in general, tends to be left aside until the appropriate time comes. Philosophers are still concerned with the problem of consciousness. Although much has been tried, a materialistic solution to this difficult problem has not yet been found. As we shall see, some scientists keep trying.

Humans are in the process of creating a limited form of intelligence, what we know as artificial intelligence. However, artificial intelligence—one or more computers—is far from conscious because it is only based on language, which is a system of symbols. That is, it works with sophisticated knowledge, but from a much more shallow level than human intelligence.

Although consciousness may have arisen from something material, which—as we shall see—is yet to be proven, consciousness itself is not something material: it is something that goes definitively beyond matter. According to Donald Hoffman—a psychologist expert in cognition, who collaborated with Crick (winner of the Nobel Prize for his discovery of DNA), trying to solve the problem of consciousness—the nature and origin of consciousness are unsolved problems. Hoffman argues that consciousness remains an enigma. According to him there is nothing that can relate conscious experience to the physical part of the brain. There is nothing in brain particles that has any relation to the qualities of consciousness, such as thought, taste, or anxiety. In short, Hoffman says brain activity is not the cause of conscious experience. There are those who argue otherwise.

As we have already seen, consciousness acquires information, orders it to create knowledge and, on the basis of this, acquires, or elaborates, wisdom. None of that is material. Nothing to do with consciousness follows the principles associated with either matter or energy, even though consciousness makes use of them. However, science seeks consciousness in two ways: one is to determine where it is, and the other is its origin, that is, how animate beings acquire consciousness.

There are scientists who believe they are close to knowing where consciousness resides within the brain. According to them, certain neural connections seem to be helpful in identifying where consciousness resides. Professor Masafumi Oizumi tells us: *"We don't yet have a conclusive answer, but a lot of empirical evidence has accumulated in the search for the minimum mechanisms sufficient for conscious experience, or neural correlations of consciousness."*

Researchers looked for what's called two-way trails. When we see or feel something, our neural networks absorb the information. But that's not enough for consciousness. Our brains need feedback, that is, to redirect the information coming from the stimuli. What was found was that bidirectionality was concentrated in the cortical and thalamic regions, that is, in the periphery of the brain and in the area called the thalamus.

Conventional neurology has studied brain activity, or information processing, which occurs in response to external stimuli. But our mind not only processes information from outside the brain, it produces the subjective experience of seeing the apple, touching it, smelling it, tasting it and imagining the tree and the place it came from; i.e., the consciousness that occurs together with neural activity.

The new discovery is that the bidirectional thalamus-cortical relationship is important in the production of consciousness. Researchers believe they're on the right track, but they're still trying to identify where consciousness actually occurs.

The second question, then, is: how do we become aware? In their recent books, *The evolution of the sensitive soul: learning and the origins of consciousness* and *Picturing the mind*, Israeli academics Simona Ginsburg and Eva Jablonka propose a precise link between evolutionism and consciousness. That is, how nature comes from the totally material to the mental, passing from the *nephesh* of the Hebrews—that is, life—to the *psyche* of the Greeks and Socrates—which in ancient times was known as the soul—but which really is what we nowadays think about as the mind, consciousness.

We ask ourselves again: Who is aware? Which animals have consciousness and to what extent do they have it? What varieties of consciousness are there? Let's start with the theory of evolution. The first principle is that all organisms descend, with modifications, from remote ancestors. The second principle is that of natural selection, which says that organisms with hereditary variations that adapt better to the environment have more offspring. It is a process of deceptive simplicity, but when applied again and again, results in organisms of increasing complexity and wonderful sophistication. Let's see, then, where should we begin to seek consciousness? At the molecular level? What happens is that genes do not cause change, but follow it. There is a change and genes adapt to it. Cultural and behavioural adaptations always precede physiological mutation. That, in turn, changes the DNA of the offspring. That is, I repeat, changes in the behaviour of the individual generate genetic changes, it does not happen the other way around. It is proven that what we do during our lives has a great impact on our genes.

Evolution presupposes a transition from non-sentient thing to a sentient being. And, of course, from a sentient being, at some point evolution takes us to a conscious, thinking being.

Here it is necessary to distinguish between several classifications of beings. According to Daniel Dennett, there is a hierarchy that ascends in order of sophistication: first there are things like sponges and plants, which evolve by natural selection, then come beings like snails or mice, which also learn from their mistakes during the life of an individual; the next step is beings learning to select between imagined actions and scenarios, such as elephants and dolphins.

Human beings, the most advanced beings, can select from possibilities represented by symbols. So, there is an approach to the issue of consciousness that admits an evolutionary transition. The inanimate becomes alive, it is followed by conscious life and the latter is followed by conscious-rational life. That is what Aristotle and St. Augustine talked about. In that chain, there are beings that have a small level of consciousness. They can perceive, feel and think from their individual point of view. But there is something else to reach the human being. A mark of evolutionary transition. An ability of the human being that does not appear in other beings. Something Ginsburg and Jablonka call it *"unlimited associative learning" (UAL)*.

Learning is a change in behaviour based on experience. Nineteenth-century thinkers believed that being able to learn by changing behaviour was a criterion of consciousness. That changed radically when Skinner's "behaviourism" appeared, which excluded terms like "consciousness" or "mind." Behaviourism—a very limited theory—ended up being rejected. However, it left a legacy: the study of associative learning. One of the things that was discovered was that learning depends on how surprising the stimulus is. A totally predictable stimulus does not require learning. According to Ginsburg and Jablonka, "unlimited associative learning" is the mark of evolutionary transition from minimal consciousness. It requires several elements, among which are the conceptualisation of objects, selective attention and active exclusion, integration through time, spontaneous activity and the existence of a goal, and most importantly: the sense of being an individual separated from others and with a stable perspective over time.

When an animal demonstrates unlimited associative learning, i.e., unrestricted learning, it means that it has the capacity for consciousness. An animal with this characteristic can exhibit complex behaviours and achieve many different goals. An animal that has unlimited associative learning can only achieve it when it is conscious. Two other elements that appear in animals with that kind of consciousness are suffering and imagination.

Imagination evolves with exploration and learning. Through experience we learn to seek what satisfies us and avoid what causes us suffering. Learning expands awareness and cognition even further, and the evolutionary cycle continues.

WELL, we've seen how scientists try to understand consciousness by studying where it seems to be in the physical world and how it seems to have originated. Something extraordinary happens in the passing of consciousness from one individual to another, from parents to children, and what astonishes neurologists is what follows.

Patrick House gives us his perspective: *"If I were asked to create, from scratch and under duress, a universal mechanism for passing consciousness from parent to child, I would probably come up with something a bit like grafting a plant. Each parent would donate a small piece of their brain and place it on some sort of growth medium... and the child would just expand..."* —he tells us—*"Instead, something much more remarkable happens in nature. An entirely new creature can grow into a fully conscious version of itself, and the entire process occurs, as if by fiat, any time a certain kind of single cell with the right mix of nucleotide sugars is kept alive for long enough. Which means that consciousness is not something passed or recycled... from one living creature to the next."*

Here, of course, what fails is that many scientists still regard consciousness as a thing that physically resides in the brain, and that the brain produces through chemical and electrical reactions. Instead, perhaps they could try to consider it as something that is not matter, shared, something that doesn't grow like a plant, nor is it transmitted in a linear way, nor is it recycled in time, but grows and is shared in space-time in a communal, social way.

Our perception of the things we can observe in nature, and their function, may change with experience. Patrick House, again, gives us the example of Marco Polo, who had the idea of a unicorn, the mythical animal, as he had learned it in medieval Europe. Seeing a rhino, Polo writes that unicorns *"are not at all as we describe them."* Seeing that the rhinoceros has a horn, Polo immediately identifies it with the pre-existing concept, but explains that, in reality, it is different from what he had imagined. Polo perceives the rhinoceros to be different, but still considers it a unicorn.

Unfortunately, things have not advanced a lot and consciousness remains one of the great mysteries for science. Historically, philosophers have wondered how the phenomena we observe give rise to conscious subjective experiences.

An Israeli physicist, Dr. Nir Lahav, and an American philosopher, Zachariah A. Neemeh, believe they have found a multidisciplinary solution to the problem of human experience. It is based on Einstein's relativistic principles. Lahav—the physicist—says conscious experience seems to be unable to arise from any physical process, which creates a huge problem for research. Neemeh, the American philosopher, believes that when we feel emotions, such as happiness, our brains establish unique patterns of neural activity. However, he says, it would be a mistake to believe that this neural activity is the emotion we call happiness. That's just neural activity that represents our happiness. Neural activity is a reflection of reality. Conscious experience, as humans define it, is not just brain activity. Scientists cannot find direct relationships between experiences and those feelings. Why, then, do we have such difficulty finding the relationship between neural activity and our observations of reality? Lahav and Neemeh say that so far two approaches to the problem have been used: naturalistic dualism, and the one that proposes that consciousness is only an illusion. Both approaches are wrong—say Lahav and Neemeh—because they assume that consciousness is an absolute property that does not depend on the observer. Again, we come across "objective reality." These two thinkers propose a conceptual argument based on mathematics: a relativistic theory of consciousness. It is neither private nor illusory, we are told, but relativistic, that is, it is based on the relationship between the observer and the observed. They recommend that philosophers studying the problem collaborate with neurologists.

Consciousness, we are told, must be researched with the same mathematical methods that physicists use to study other relativistic phenomena. Totally agree. On a macrocosmic level, we are told that time begins with the famous Big Bang, although that is not verifiable either. That is a deduction from science. The non-existence of time before the Big Bang is something yet to be proven. Now they say that things that have been seen through the James Webb telescope contradict the possibility of the existence of the Big Bang (in terms of having been an event; the universe is known to expand), and also that there are galaxies that are much more than fourteen billion years old. That—it was believed—was the age of the universe.

On a microcosmic level—that is, individual—we begin to have an idea of time when we acquire reason, not before. We know that we perceive it in different ways during the course of life, and we intuit that, for the individual, time ends at death. As I explained at the beginning, there are many things we take for granted with respect to time, although we already know that it is not absolute, as Newton claimed. We also said that some scientists deny the existence of time. I know, they do it at the quantum level. But it could also be said that, at our level, time, finitude, and death exist only in human consciousness. Animals, which have no consciousness, or a limited one, have no idea that they will inevitably die, nor do they know time. They live a present that is changing without them realising it. In that sense, a crow is immortal. But we are not immortal. We have consciousness. In reality, we are part of this great Conscience that Christians called the Holy Spirit.

In that respect, Bertrand Russell tells us: *"The individual existence of a human being has to be like a river—at first, barely contained by the banks, rushing passionately through the rocks and over waterfalls. Gradually, the river becomes wider, the banks move away, the water flows more slowly. In the end, with no visible limit, it ends up mixing with the sea and, without the slightest pain, loses its individual identity."*

Consciousness—which in the most basic sense of the word is to feel inner and outer existence—is also something shared, as we shall see, at a different level. Human beings share it because we are social beings. We share consciousness, but we also feel like individuals.

We have an individual identity. As we said, a tiger may not need an identity, because it lives in solitude. We need one, just as we need a name, for others to identify us. What we don't know is whether our "qualities"—subjective sensory qualities—that is, perceptions of colours, smells, or sounds, are the same as those perceived by others.

The most difficult problem concerning consciousness is how and why we experience natural phenomena the way we do. The knowledge, memories, sensations, thoughts and dreams we experience determine our individual identity. Collective consciousness resides in the beliefs, attitudes and ideas that unite us to the rest of the society in which we live, that we share with that society.

We saw that Wallace did not fully share with Darwin the belief that the strongest individual is the one who survives. In our current human society, to some extent, that principle does not work. We have developed a society in which the state is given the monopoly of violence and which, ideally, protects the individual.

At present, humans as a society, have sufficient resources and lack competitors, or predators of other species, that can put us in a situation of danger. On the contrary, we are so successful that we have reduced the other species to such an extent that many have become extinct. Also, our science advances in the fight against disease.

Consciousness gives us the power to imagine, to visualise the future (to imagine is to create an image). We act mostly on the basis of imagination. In that way we can also create, which is the ultimate human power. We can do it alone or by combining our consciousness with those of other human beings, whom we convince, or who convince us, to work together. That combined consciousness, which at one time allowed us to hunt animals much more powerful than ourselves—and now allows us to build a thousand things, from skyscrapers, bridges, satellites, to aircraft, or computers—can act on the basis of the principle that "the whole is greater than the sum of the parts."

Schrödinger believed in a universal consciousness. In his book *Mind and Matter*, he said that *"The total number of minds that exist in the universe is one."* By which he meant that there was a single consciousness. I find many parallels with his philosophical position. I always had great coincidences with his thought without having had any notion of doing so (although I had heard a lot about Schrödinger, I had not read anything written by him). In the book, Schrödinger explains very clearly what he believes about objective reality: *"I argue that this is a kind of simplification that we adopt to solve the infinitely complicated problem of nature. Without knowing it, and not being rigorous in a systematic way, we exclude the Topic of Cognition, placing it outside the field of nature that we try to understand. We place our own person in the place of an observer who does not belong to a world that thus becomes an objective world."* Science, then, finds it very difficult to deal with time according to its normal canons. As I shall explain later, Western science and Christianity are part of the same paradigm in which human beings are separated from the rest of the universe.

In the book, Schrödinger mentions two antinomies regarding objective reality: the first is that our image of the world is "colourless, cold, and mute," because colour and sound, and sensations of cold and heat are immediate sensations that we cannot reproduce in reality by taking our mind out of it. We can talk about degrees of temperature, but not about the sensation itself. The second antinomy is that science cannot find the place where mind acts on matter or vice versa. In doing so he quotes Sir George Sherrington, but more than anything he relies on Spinoza's *Theory of Attributes*: "*The body cannot determine the* mind *to* thinking, and the mind *cannot determine the* body *to* motion, to rest, or *to anything else (if any)."*

Schrödinger was totally convinced that metaphysics did not follow physics, but preceded it. Metaphysics is not a deductive discipline but rather a speculative one.

I understand that Schrödinger's thought was greatly influenced by Indian philosophy. What I don't know is whether the link with Eastern philosophy began by discovering quantum mechanics and noticing that the observer exerts influence on the object—or if the opposite happened—and Eastern philosophy led him to that discovery. Some claim that it was the latter. He said, *"We cannot declare anything about a given natural object (or physical system) without contacting it. That contact is a real physical interaction. Even if only "the object is looked at", it must have received a ray of light and reflect it on the eye, or on some observation instrument. That means the observation affects the object. No knowledge of an object can be gained while leaving it strictly isolated".*

I HAVE INCLUDED as much information as I could on the different perspectives we have on consciousness. Now let me tell you why I see mistakes in much of what has been said about consciousness thus far.

Let us divide consciousness into two elements: 1) basic, animal, consciousness, which we share with all other sentient animals (I will call this one *nephesh* for short, and later will explain why); 2) high, human, consciousness, which allows us to communicate, to have sophisticated thinking, to speak languages, to create, to invent, to collaborate on large projects, etc. (I will call this one *psyche*, which I think needs no further explanation).

We are born with *nephesh*. That is evident: we can perceive. That involves visual, auditive, olfactory, gustatory and tactile perceptions. In different ways, all sentient animals share, in varying degrees, that way of perceiving reality.

We acquire *psyche* through a complicated process. It involves parental, cultural and collective participation. This happens for a reason. Later I will explain why I believe it is the way it is.

The result is that both elements of our consciousness overlap, but are not continuous. They are discrete elements. Studying *nephesh* does not necessarily mean understanding the workings of *psyche*, and vice versa. The dichotomy has clear and real consequences. Thus far, Western science does not appear to have understood this fact. *Psyche* includes components (or extensions, as St Augustine would have called them) whose existence cannot be inferred from a study of *nephesh* or of the history of *nephesh*. Understanding *psyche* requires a new approach.

In the Conclusion and Afterword I'll try to explain my views on consciousness in greater detail.

CHAPTER 2

TIME

"In the wordless beginning, space-time itself was squeezed and squeezed into a little ball in which everything gathered: what we call the singularity. Even if sound had existed then — it didn't exist, of course, because sound is made of matter — everything would have existed at the same time. Infinite amounts of every possible note would have been playing at the same time — the antithesis of music. Time was only born because that single point of totality stretched into a line and suddenly there was continuity. Suddenly, one moment began to distinguish itself from the other — the strange gift of entropy, which makes possible the existence of melody and rhythm, chords and harmonies."

- Maria Popova

What Maria Popova says is beautiful. I don't agree with her, but it takes the reader to an imaginary place where sound is made of matter and time is born. And it is fascinating. That is what many people want to believe. But if it is about believing, I do not believe sound is made of matter. Sound waves are not matter. I believe sound exists within a gooey, sticky present without which it would not be able to exist. I believe sound exists because there is a listener, without whom it would not exist. I believe that time was not born then. I believe what probably began then, if ever, was change. I believe time evolved with human consciousness.

We've been measuring time since "time immemorial", if that makes any sense. Perhaps I should explain: a recent article by Bennett Bacon *et al*, *An Upper Palaeolithic Proto-Writing System and Phenological Calendar*, published by Cambridge University Press, describes research on paintings, conducted in hundreds of European caves, and on engravings of bones.

The objects of the study were depictions of animals (prey) by *Homo Sapiens* some 37,000 years ago. For a long time, the depictions were believed to be art. It has become evident that they were not art. They were mnemonic and notational devices. Maybe I would not call them proto-writing, but the study found that frequently occurring signs, like dots, lines and "Ys", paired with figures of animals, were meant to carry meaning. The symbols signified months and seasons; they were part of a calendar beginning in spring and recording lunar months. The "Y" sign was indicative of parturition of the particular animal next to the notation.

What is fascinating about the research is that it demonstrates, among other things, that human beings, since that early stage, had been measuring lunar months (time) for hunting purposes. That is, that time (change and the repetition of seasons) had preoccupied humans for strategic purposes since the beginnings of human consciousness. The finding leads me to believe that there is a close association between the evolution of human consciousness and the concept of time.

In Spanish we have a very good word: "*siglo*". In English it is called "century". But the two don't mean exactly the same. Century is a word derived from one hundred—from Latin, "*centum*". That's a precise number, whereas "*siglo*" was not originally a precise term. And what does "*siglo*" mean? Where does the word come from? It comes from "*saeculum*"—diminutive of "*saecum*" (an era)—actually an Etruscan word that Romans adapted into Latin. It seems to mean "a short time, a few years," or perhaps "the memory of someone who is still alive": a lifetime. In English there is a very similar meaning: "within living memory". The Romans sometimes assigned it ninety years; others, up to one hundred and ten. Better to give an example to clarify what I mean. In his short story *The Witness*, Borges tells of an old man who is dying in a stable in England. The bells toll, calling for prayer. England is already Christian. But that old man remembers the god Woden, whom he had seen as a child; a crude wooden idol with Roman coins, to which horses, dogs and prisoners were sacrificed. When the old man dies, the last person who has seen the pagan god will die. When he dies, humanity will have lost something. Perhaps Borges did not remember, or was improbably unaware of, the hidden meaning of the term *"siglo"*, but what he says in the story is just like what the Etruscans were trying to say. It's a bit like when the last person to speak a language dies. At that moment, the language also dies; humanity becomes poorer. I think we all know, or at least intuit that, part of our collective consciousness is also a collective memory, because we think about the present, but we also know the past and think about it. Either way, we know that Julius Caesar lived. Brutus and the other conspirators killed him. Let us forget what Shakespeare wrote: there were real people—like the one who is reading or like me—who saw it and who heard Mark Antony's eulogy after the murder. Tacitus and Plutarch wrote about him. In Spanish we have adapted a word like *"siglo"*, with its flexible meaning of time and we have made it an exact, scientific word, like "century", in English. Now it means exactly one hundred years. How long it has taken for our planet to circle one hundred times around the Sun.

The most incredible thing is that now we are all interested in knowing what "time" really is. After Einstein, the concept became "space-time", although we will see that his new idea, apart from adding an element, is much more complex than that.

Throughout history, and really because we needed to, we came up with increasingly sophisticated ways of measuring time. The Egyptians and Sumerians had already divided the year into twelve months of thirty days. They also divided the day into twenty-four hours.

Of course, even if we assume that the way we measure time is accurate, that is also not true. Our concept of time is very limited and we have been inventing it little by little. As we have been inventing how to measure it.

For the ancient Hebrews, the day began when it was possible to distinguish a black thread from a white one. The days ranged from dawn to dusk. The idea had an indisputable logic. As it is indisputable that the Sun rises every day in the East and sets in the West: that happens three hundred and sixty-five times a year. That is, our planet rotates on itself those three hundred and sixty-five times as it circles the Sun. So far, so good. Our scientists have discovered that and we know it's a fact. A fact that we can even measure. Very good.

There are many other things that seem true but are totally arbitrary. The only certainty about them is that we are used to them.

A year has twelve months. A year has fifty-two weeks. This is the year 2023. A week has seven days. Really? Are those facts or inventions? Well, the way we divide time is—by the way—an invention.

We have naturalised the divisions of time in such a way that it is already instinctive. My grandfather had a beautiful pocket watch that he never used; he knew the time without looking at it, almost exactly. "What time is it?" we would ask. He would answer with certainty "It's 10.35". Then, he would tak the watch out of his pocket and it would be 10.36. "How do you do it? What's the catch?" we would ask. "There's no trick," he would tell us. "I'm old." Now the same thing happens to me. And I understand.

Let's see how some concepts slowly evolved and now we assume they are true just because of that.

In the ancient East, in China and Japan, the day was divided into twelve sections of two hours each, which depended on the signs of the zodiac. They varied according to the time of sunrise and sunset.

A Chinese dignitary named Su Song invented a kind of clock, with a tower about twelve metres high. The mechanism worked thanks to a stream of water. As it turned, the clock showed the time to the citizens of Kaifeng and at the same time spun a celestial bronze globe with the positions of the stars. In a certain era, time was also measured in China with a small labyrinth full of incense that burned as the hours passed.

In Japan, during the Edo period (after the Portuguese and Spanish tried to introduce Christianity) the day was divided into six parts, between dawn and sunset. The time allocated to those periods varied according to the season, either at night or during the day. The Japanese of those years also measured time following the bells of castles and temples as they tolled.

In the West—and now in most of the civilised world—we use a calendar that we inherited from the Romans. The matter gets quite complicated, but let's just say that originally their year lasted ten months.

The first Roman calendar began in March (after Mars, the god of war). April was the month of Aphrodite, the goddess of love. Then came May, for Maia, a goddess of fertility. And then June, named after Juno, another important goddess who protected Roman society.

After that fourth month, the Romans got bored and numbered all the following months: fifth (Quintilis), sixth (Sextilis), seventh (Septembris), eighth (Octobris), ninth (Novembris) and the year ended in Decembris, which means tenth month, as you may have already guessed, although it ended up being our twelfth month.

What happened is that January (Januarius), named after Janus, god of doors and windows, and February (Februarius, month of purifications, were winter months and did not count. As I explained, the matter is complicated, and much less scientific than we think.

But it gets more difficult as, finally, January and February were placed before the other months, which made the calendar twelve months long. By that time, after Julius Caesar's death, Mark Antony renamed Quintilis, Julius, after Caesar. Later, when Octavian became the first emperor —that is, Caesar Augustus—Sextilis became August. These, then, are the twelve months we have now in the West. Named in a totally arbitrary way by the Romans. There are many other complications regarding the way we measure time, including the days of the week, although it is better to leave the calendar right here. Believe me, today is not Thursday.

In 1582 the Gregorian calendar was introduced. Until then the Julian calendar had been used. The difference was that, that year, Thursday, October 4, became Friday, October 15, with the consequent loss of ten days. That corrected an error that the Julian calendar had had, in which the year was eleven minutes longer than it had to be.

And the year now, is it 2023? Yes, it's 2023, for us. Well, actually 2023 is the number of years we believe have passed since the birth of a prophet, part of the Trinity, or God the Son, according to Christians. And I specifically say "believe" because in reality, we are not sure that that is the number of years since the birth of Jesus of Nazareth.

We say that planet Earth is 4.5 billion years old. And we measure the time of its existence in years because it's the only way we can begin to understand that amount of time. That is: a year is the time it takes the Earth to orbit the Sun. But 4.5 billion years ago—if there is anyone who can understand a figure of that magnitude—neither the Earth nor the Sun existed.

Something similar happens with distances measured in light years. A light year is the distance that light travels in a year, that is, 9,460,730,472.581 kilometres—that is, almost 9 and a half billion kilometres. No one, but no one, can imagine that distance. Although Einstein said he could imagine what it would be like to travel at the speed of light.

In Europe, among the many devices that were used to measure time, we may have begun with the clepsydra—κλέπτειν (kleptein, to steal) and ὕδωρ (hidor, water)—from the Greek (a holed bowl that was filled with water and emptied within a certain period), which the Romans later adopted (and adapted) to measure the speeches of their senators.

Although I speak of the Greeks and its Greek name, the clepsydra is actually Egyptian. In 1500 BC, there was an inscription on the tomb of an official named Amenemhet that he had invented a type of clock that measured time during the night. The invention was an alabaster bowl with a hole and twelve marks inside bearing the names of the Egyptian months. The marks measured the length of day and night during the different seasons. Experiments have been done suggesting that the shape allowed a constant flow by which time was measured with a tolerance of between ten and fifteen minutes per night. Sufficient accuracy for the time.

In medieval monasteries, the day was divided into canonical hours—and in some of them it still is—; the hours were marked by ringing the bells and during those times they prayed or sang, or both. The liturgy of the hours, also applicable to all believers, began with the *matins*, more or less at dawn, and passed through the *lauds, prima, tercia, sexta* and *nona*, until it reached *vespers* and the *completas*, which was about nine o'clock, that is, the hour of rest.

A primitive alarm clock that was sometimes used was a candle with perpendicular nails at a certain height. As the candle burned and the wax melted and, at the desired moment, the nails fell on a metal plate and made a noise that woke the person up.

Once the Europeans finally decided that the day had twenty-four hours, sundials were invented, and they were everywhere during Middle Ages, until mechanical clocks were first seen in the fourteenth century, on church towers and on the façades of town halls and other public buildings.

In the same way that happens now with electronic elements, necessity made mechanical watches become smaller and smaller. It is said that the first pocket watch was invented in the seventeenth century. It is still debated whether the first wristwatch appeared in 1812, created by Breguet, or whether it already existed in the sixteenth century.

At present there are many brands and models of wristwatches, analog and electronic. There are watches for running, for sailing, for skydiving and diving. The Apple brand has sold more than one hundred million *Apple Watches* in the world.

We now know that time passes faster the closer you are to the centre of the Earth. So, according to a clock that is on the desk, time passes faster than that of a clock that is on the floor. We don't just know this. We can also measure it. There are now laboratory clocks accurate enough to measure the difference. A century before we could see it, Einstein knew that this difference existed. Scientific thinking is exactly that, being able to understand something before you have observed it. What I question, and we will get to that, is what good is this time difference in the experiences of our daily lives.

Some will say that, in a ski race, for example, a tenth of a second can mean a first prize and a lot of money. The same thing happens when a concert pianist misses a fragment of a second and the feeling of the melody changes. As for space—and here I digress a little—very small differences may make an architectural project fail. An error of one millimetre can cause a machine to not work. Painter Josef

Albers, a famous perfectionist, once told his friend and client, actor Maximilian Schell: *"Today is one of the most glorious days of my life. I had drawn a picture twenty-seven years ago and knew there was something wrong. I just figured out what it was. The line above was half an inch below where it needed to be. I just fixed it and now the drawing is perfect."* Space and time are important to us in many ways that sometimes we don't even think about.

But science and philosophy teach us to doubt everything. We have to doubt even our reality and our experiences. As we said before, according to science our senses are not reliable. There are certain things that may seem real to us, but appearances are deceiving. We know that sometimes things that appear very obvious to us are just an illusion, part of a preconception.

We have learned a lot about time. We said that, according to Einstein, time is relative, not absolute. It goes forward. As we have already seen, the universe has an "arrow" of time that flows towards an entropy, towards an increasing disorder. According to quantum mechanics, when we reach an infinitesimal, quantum level, it seems that time is not necessary. It would be important to understand the way in which time might not exist.

Aristotle says that time consists of two parts, since the present—the now—is not a part (I still don't understand this; perhaps what he meant was that the present ever changing, that is, something that becomes). He also says that time is not absolute but relative. The future is going to be at some point, but it still isn't, and the past is no longer there. Therefore, for him, the existence of time is somewhat doubtful.

Aristotle asserts that there is a relationship between time and change, but that they are not the same thing. Time is not change itself. After much rumination, he comes to the conclusion that time equals the amount of change. But he also says that it is not a type of measure but a type of order, as Rovelli says. He then comes to the following conclusion: time depends on the soul, but it can exist without the soul because it can be numbered and numbers are eternal. I do not agree with that last part, as numbers only exist within human consciousness. Numbers are one more creation of the mind to be able to understand reality. But he also says that time can only be counted if there is someone to tell it. In the end, we agree: he says that without human consciousness there is no time.

WE ALREADY EXPLAINED THAT, according to physics, if something can be measured, quantified, defined mathematically, is an observable quantity on which other observable variables depend, that something exists.

To be real, something has to meet all those conditions. In physics, if something is not possible, it is called a "pathology." Maybe time is pathological? Maybe it's impossible? Well, it meets all the conditions we said before to be possible. I think it would have to be real. The problem is that the answers to all those questions are relative. They depend on where you are or if you are moving or if you are in the same place. That's what Einstein proved. Time is real, although it wouldn't seem to be real in an objective way. How?

Let's clarify this. According to the theory of relativity, time is not pathological. It is only relative. That means that if we put together the person who is on the train with the person who's still at the station, their times will be different, but if we do it again and again, with different people, the results will be consistent. The results are going to be predictable.

So, the idea that "time is relative" does not prove that time doesn't exist, or at least that's what it seems to me.

Let's look at it from another perspective.

We said that there are differences between Einstein's theory of relativity and the discoveries of physics regarding the quantum properties of space and time. There is still no generally accepted theory of quantum gravity, which would link relativity with quantum mechanics.

However, the diversity of opinions regarding the nature of time has been decreasing. Many now understand that the temporal part of the theory of relativity disappears the moment we consider the quantum perspective, that is, from the moment we consider the world at a minuscule level.

And what has quantum mechanics discovered with respect to time? Well, three fundamental characteristics: its granularity, its indeterminacy, and its relationship with other physical variables.

Granularity is something characteristic of quantum mechanics: "quanta" are elementary granules. There is a scale, called the "Planck scale," that measures the tiniest chance of time in the gravitational field. The smallest chance of measuring time is called "Planck time." It is $[5.319124 \times]$ 10^{-44} of a second. That is, a hundred millionths of a trillion, a trillion of a trillion of a second. A figure more than incomprehensible. Planck's time, then, cannot be measured. No current clock can. As we said, according to physics, if you cannot measure it, you lose a condition for the existence of something. If something is not measurable, it does not exist. On such a minuscule scale, the notion of time ceases to be valid. Time does not exist. Well, it does to a certain extent.

Carlo Rovelli, an eminent Italian physicist, writes in his book "*L'ordine del tempo*" that granularity is a universal characteristic: "*Perhaps the rivers of ink that have been spent talking about the nature of the 'continuous' through the centuries, from Aristotle to Heidegger, have been wasted. Continuity is just a mathematical technique to approximate things of very fine granularity. The world is subtly discrete, not continuous. The good Lord did not design the world with continuous lines: he did it with a light hand, he sketched it with dots, like a painting by Georges Seurat.*" Although we may not agree, it must be said that the man explains it in a brilliant way.

Planck time has its equivalent in space: "Planck space", which is equivalent to 10^{-33} of a centimetre, one millionth of a trillion of a trillion of a trillion millimetres. Things like that are dealt with by quantum mechanics. Indeterminacy is another characteristic that is discovered at that level. You can't predict when an electron is going to appear. Between one appearance and the other, the electron does not have a certain position. It disperses in a cloud of probability. This is called an "overlap." As quantum mechanics explains, space-time fluctuates and can overlap in different configurations. Again, Rovelli tells us: *"There is no single time. There is a different duration for each trajectory; and time passes at different rates depending on the place and speed. It is not directional: in the elementary equations of the world, the difference between past and future does not exist; its orientation is only a contingent aspect that appears when we look at things and do not observe the detail. In this blurred view, the universe's past is in a curiously "particular" state. The notion of 'present' does not work: in the vast universe there is nothing we can reasonably call 'present'."* ... *"[Time] jumps, fluctuates, materializes only by interacting and, below a minimal scale cannot be found... So, after all this, what's left of time?"*

There are things that are real on a physical level, work is real; entropy is real; thermodynamics is also physically real. However, although it cannot be measured at the quantum level, time is a measurable, observable and quantifiable entity at our level, at the classical level. We know, then, that it is real at our level. But there are things we still don't understand. For example, we do not know the causes of what is called the "arrow" of time. We see that time flows forward and not backward; we become aware of the passage of time and are subject to its movement, like all physical objects. Even if the entropy of the system remains constant, increases rapidly, increases slowly, or decreases by adding energy to the system, the "arrow" never stops or goes backwards. So, we explained, at our level, time is real, although it may not be fundamental (we'll see). At present, physics considers entropy to be a derived quantity and treats time as fundamental. Mathematically, however, if we say that entropy is a fundamental quantity, time becomes an emergent quantity and behaves as such. This is a bit difficult. I'm sure Hawking understood that. Time appears to be an integral part of the universe. The problem lies in how we perceive it and our ignorance of the way it works. Scientists tell us that, since the hot Big Bang, time progresses in a certain sense for all observers; that, in the same way, there is an "arrow of time", that the universe expands according to the laws of thermodynamics, and that entropy grows. If they say so, so it is... or is it? Time passes and objects move. Things change.

Scientists believe things are going to continue like this. They also believe that the arrow of thermodynamics and that of time may be related. There is a very important symmetry in physics which is the symmetry of time reversal. If time were to go both forward and backward, the laws of physics would hold. Gravity, electromagnetic and nuclear force remain identical regardless of the direction of time. Is that so?

Entropy is the extent to which a configuration of particles, once undone, is likely to be able to reconfigure. If you beat an egg, you can't go back to have the whole egg again. Theoretically, it could be done, but the probability is almost zero. In the case of the beaten egg, the probability is infinitesimal. The particles in any system have a finite number of possible configurations.

In physics, currently, a couple of things seem to be certain: the entropy of the universe always grows, and time always goes in the same direction. Is that correct?

Much has been said about the famous experiment that proves that a particle passes through two or more slits at the same time. That happens when the particle is observed.

When a quantum particle approaches a barrier, it may bounce, or perhaps it will break through the barrier. The evolution of the particle is only determined by observing it. The interpretation of the wave function only applies to the unmeasured system. Once the trajectory has been determined (once it has been observed), the behaviour of the "past" is the same as that of the classical level, our non-infinitesimal level. Now, even the existence of the wave function is in question.

One version of quantum mechanics says that there is a classical world and a quantum world. What divides one world from the other is the moment when things become defined, when indefinite things become defined in the quantum world. That moment is the present, the now. I think that's a super complicated way to explain reality.

Time may or may not be fundamental, and our perception of the "arrow" of time may or may not be related to thermodynamics. The point is that, at our level, at the classical level, time can be measured, observed and quantified.

Physics appears to suggest that everything has to do with the "relativity of relativity", that is, taken to its maximum expression, relativity becomes relative. When any analysis is performed at the micro-cosmic level—quantum—relativity has no applicability. The only thing I understand from all this is that the two theories do not coincide and that it is necessary to discover a "quantum gravity" to make them compatible.

All the laws of physics we know are symmetrical with respect to time. All those related to motion, gravitation, electromagnetism, are completely reversible in time.

As we said, if one beats an egg, drops a cup on the floor, or has a car collision, one creates a situation to which a thermodynamic "arrow" of time is applied. These events are not reversible, or it is highly improbable that they could be reversed, so we perceive time as unidirectional. But, do we really perceive time? Or do we perceive change?

And then? What do physicists mean when they say that time doesn't exist? Why do they say that? Physics does not include that coffee makers exist, that there are people, that there are buses, but nevertheless accepts that coffee makers, people and buses exist. Physicists accept that coffee makers, for example, "emerge" at a higher level than physics describes. If time is a fundamental property of the universe, perhaps it also emerges. The thing is, it's easy to understand that a coffee maker is made of particles. But time... what is it made of? Our divisions of time, as we have already seen, are rather arbitrary inventions.

However, in our lives we need time. Our lives *are* time. We live now, learn from the past, and plan for the future. Are we the only species who does that?

If time didn't exist, why would we have to get out of bed?

Let's put it this way, if physics needs to say that time doesn't exist to explain something, there it is. There is a reason why it will. Science also said that reality is objective to somebobdy like Schrödinger. There are many things that are very debatable. As I said before, we'll look at all of this in greater detail. In any case, not all scientists agree on those things.

Rovelli says there is nothing mysterious about the absence of time in quantum gravity. What happens is that at the fundamental level there is no variable for time. Time, at that level—as we have already seen, is what they call "Planck time"—so minuscule that it is negligible. Fantastic, that's as far as I understand.

But it also says something that I found very interesting. As an admission that time is not totally absent in quantum mechanics: when you change the position of a molecule, so does the state of the molecule. The same thing happens when changing the speed: there is another change in the state. But if you change the velocity first and the position later, the state of the molecule changes in a different way. That in quantum mechanics is called "noncommutativity" (I looked for "incommutativity" but it doesn't seem to exist). In other words, the order of the changes cannot be altered without altering, also, the state of the molecule. Rovelli says *"La non-conmutatività determina un ordine, e quindi un germe di temporalità"*. At the quantum level there is a germ of time! It seems physics has a bit of poetry to it. But, is it true? Is the germ a germ of time?

Perhaps that is only putting a bit of feeling into something that is simply without any feeling at all, like physics. Of course, science—which I respect very much—can tell us, through its daughter, technology, how fast an airplane is going, its weight and mass, or why we can't have our cell phone on during the flight. What it can't tell us is whether it would be good for us to go see aunt Eugenia, who is very depressed. Nor can it tell us if we should yell at the neighbour to silence his dog. On the contrary, philosophy, art, poetry, music can give us some idea of what is right and what is wrong, or what makes us feel good or bad.

In the previous chapter, I discussed consciousness, and then explained my views. In this chapter, I have given all the information I could on time and the history of time as perceived through the Western mind. Now, these are my views on time, which I will expound in detail later.

WE HAVE SEEN, and I have hinted at the beginning of this chapter, that time has preoccupied human beings throughout history, and even during some prehistory, as explained. But that preoccupation does not extend to other sentient animals. Why, if we share some form of consciousness with them?

The answer lies in the discreteness of the two consciousnesses. During its evolution, *psyche* appears to have introduced the notion of time. Again, why is that? Because human beings, in using their unlimited capacity for learning from experience, developed memory and imagination. We have long term memory and we can apply strategy to our actions. Time is the tool we need to measure the long term memory and the expectations we have developed.

As individuals, we learn about time from our parents, from the collective, from the culture, from the language they use. Time is important to us as human beings. The way it develops in human minds is individual. Time is taught again and again to every child. It does not come as an instinct. We learn it.

Time does not disappear inside black holes. Time is not granular. Time may or may not bend in space, it depends on who's studying it, if anybody is. Physicists can speculate all they want, but time is only a human device, like numbers and identity. That is clear to me.

Time, like dreams, is an ethereal creation of the mind. The difference is that dreams are individual, whereas time is a tool used by the collective to have a clearer idea of when memories occurred and when expectations will occur, if they become a reality. For all we know, our cave ancestors only thought in terms of seasons and the parturition of their prey. That was all they needed.

My assumption is that Alexander Selkirk—the real Robinson Crusoe—didn't need to measure time in order to survive on his island, except for knowing when days and nights and seasons would occur, and, if he did measure time further than that, it was just as a remnant of his life within human society.

The numbers we need to measure time exist only as far as we need them. Some cultures and some languages use numbers and time to a very limited extent.

In summary, time exists; but it exists only within *psyche*. What happens when there is no *psyche* is just change. Animals know that. Aristotle would have agreed with that. We did not discover time. We invented it.

CHAPTER 3

OBJECTIVE REALITY & QUANTUM PHYSICS (AN OPTIONAL CHAPTER)

> *"25. ¶ And God made the beast of the earth after his kinde,*
> *and cattell after their kinde, and euery thing that*
> *creepeth vpon earth, afteer his kinde:*
> *and God saw that it was good.*
> *26. ¶ And God said, *Let vs make man in our Image,*
> *after our likenesse: and let them haue dominion ouer the*
> *fish of the sea, and ouer the foule of the aire,*
> *and ouer the cattell, and ouer all the earth,*
> *and ouer euery creeping thing that creepeth vpon the earth."*
> - Holy Bible, King James Version

The Bible? A quotation from the Bible? A whole thingame on quantum physics? Please! ... Well, it's not as bad as it looks, it's optional. And we need a bit of patience.

I wanted to add this chapter because it I believe it would provide a bit of a historical perspective and a better understanding of how Western science views consciousness and time.

I repeat, according to science, if something can be measured, if it can be quantified, if it can be defined mathematically, if it is an observable quantity on which other observable quantities depend, then that thing exists. As we have seen, with time, it's a little more complicated than all that. We'll see.

Let's analyse this: in the field of physics some suggest that there is a possibility that time doesn't exist. This seems to be the type of nonsense that science sometimes comes up with. Science often makes mistakes while learning. That's part of the process.

Physics is in crisis. For about a century the discipline has been divided into two equally successful theories: the theory of relativity and quantum mechanics. Both of them reject Newtonian mechanics, based on objective reality, something the rest of science needs.

As I think I have already explained, quantum mechanics describes how things work in the infinitesimal world of particles. On the other hand, the theory of relativity, introduced by Einstein, explains gravitation (in our world, gravity) and the motion of objects at a higher level. The two make sense, but so far, they have not been able to agree with each other. So, we need a unique theory of "quantum gravity" that fuses the theory of relativity with quantum mechanics. I think that's what some time ago would have been called a "unified field theory," or at least that's how my dad explained it to me when I was a kid. The Holy Grail of physicists. Basically, a theory of quantum gravity would be of great importance to science because it would unite our views of reality, classical and subatomic, and allow gravity to be incorporated mathematically to achieve an "integral theory."

Quantum gravity would solve how gravity works at the particle level. Producing such a theory is extremely difficult. Two ways have been tried: one is "string theory," which replaces particles with strings that vibrate in eleven dimensions. Without going into details, that seems not to have gone well.

The newest is called "loop quantum gravity." It proposes that space is composed of discrete chunks, or "loops."

The original idea was to write something more or less easy to understand about consciousness and time: to make a kind of analysis and include a number of different perspectives on the subject; to clarify the matter, if it is possible to clarify things that we all think we have very clear. Although, as I explained before, I also try to question some of the things that many of us take for granted.

Needless to say, apart from scientists and philosophers, the rest of us human beings also have our doubts regarding consciousness and time.

Let's see how Western knowledge has progressed, why the issues of consciousness and time are so difficult for science, and why I believe that science has reached a decisive crossroads in the development of its knowledge on the subject.

I'm going to go back a long way. Some may think the idea is far-fetched, but I think it's worth developing it simply and deciding whether the reasoning makes sense.

I depart from a fundamental point: I believe that Christianity and science are opposite parts of the same paradigm. Something very Western. Here, many are going to say no, they are totally different things and what you're saying is crazy. OK, let's look at the grounds for my assertion.

The greatest developments in science and technology have historically occurred in the West. Why? My analysis is as follows, and it begins a little before Christianity. I go on to describe a concatenation of events that lead to the concept of "objective reality", indispensable for the development of Western science and technology, and a source of stability and certainty in their discoveries. Objective reality—a basic concept—seems to have surpassed the limit of its usefulness. Quantum mechanics questions objective reality. Let's see how it originated and why it's not viable in all aspects of science.

1) Ancient Hebrews, who did not consider human beings as separate entities from the rest of the animals, had the concept of *nephesh*, the breath of life that God gives to animals and human beings. At the end of life, the being returns to dust.

2) There came a time when thinkers added something new to the existence of life: what was needed was an explanation for consciousness. The Greeks invented the term *psyche* —the soul—although today we no longer speak of soul, but directly of consciousness. Plato came to the conclusion that the soul was immortal, which has a certain logic when we realise that our knowledge and our thoughts are transmitted to contemporaries—that is, in space—and pass from one generation to the next—that is, in time—which follows indefinitely. Thus, the consciousness of the human being passes exponentially from one individual to many others.

3) Saul of Tarsus—a Jew born in Cilicia, part of the Hellenistic diaspora—later known as St. Paul, who spoke Greek, Latin, Hebrew, and quite possibly Aramaic, was aware of the most important ideas the Greeks had. When Jesus of Nazareth died, Saul took charge of the sect and imposed the idea that the human being, apart from having been created as a separate entity from the rest of the animals, had an individual and immortal soul that went to heaven with God. The human individual, then, was separate from the rest of the universe. From being a peripheral sect of Judaism, Christianity spread and, with Constantine, it became the religion of the Roman Empire, that is, of most of the known world.

4) The Church then grows to become the most important institution in the life of the West and governs the life of individuals. The Renaissance encourages the validity of the individual as an observer. Arts and sciences mature in Florence. In Florence itself, Girolamo Savonarola, a Dominican monk, rebels unsuccessfully against the Church and, in 1498, is executed. In 1517, Martin Luther—another

monk, this time an Augustinian, who lived in Wittenberg, in what is now Germany—rebels again against the Church and declares that the individual can relate directly to God through the Bible. Western individualism reaches its peak.

5) Isaac Newton, from Cambridge, in England, experiments with alchemy, astrology, mathematics and theology, to prove the existence of a force called gravitation, which governs the movements of the stars. Newton imposes his ideas on the mechanics of celestial bodies, and establishes the logical process of research. Human beings can study the universe as something immutable that surrounds them, but they are totally separate from it. Science flourishes. Objective reality has been reached.

So far, the evolution of the concept of objective reality in the West.

In nineteenth-century England, Charles Darwin—after having travelled around the world studying the geology, botany, and development of the various species of animals in South America and Oceania—elaborates in great detail a theory originally conceived by his grandfather, Dr. Erasmus Darwin: the origin [and evolution] of species. The study is meticulous and scientific, deals exclusively with material aspects of species, and is grounded on the canons of the period.

Completely independently, another multifaceted English scholar, Alfred Russel Wallace, also after a years-long trip around the world, comes to the same conclusion at about the same time: species evolve. However, Wallace doubts that an entirely material explanation of human consciousness can be given. The origin of spirit, he says, is not explainable through matter. An agreement is reached that the theory of the origin of species remains as something exclusive to Darwin, and he is awarded distinctions and scientific glory.

In Western countries, science and technology are advancing overwhelmingly. Meanwhile, in the East, knowledge is going down a totally opposite path and the conclusions of Eastern philosophy do not seem to lead to the same kind of advance. The individual human is part of society and nature, not separate from the rest of reality.

Going back to physics, also in the nineteenth century, Pierre-Simon Laplace predicts that, if the positions and velocities of all the particles in the universe were known, their behavior could be calculated in the future.

At that time, it was thought that all hot bodies emitted radiation, and lost energy by means of radio waves, infrared, light, ultraviolet, X-rays, and gamma rays, *in equal quantities*. That is not the case. If it were, the entire universe would have the same temperature.

In response to Laplace, Max Planck states, in the early twentieth century, that energy is lost in "quanta"—discrete, very small particles—and that energy is lost in greater quantities in ultraviolet and X-ray light than in visible and infrared light.

In 1905, Einstein proposed the Theory of Relativity, which questioned Newtonian mechanics and revolutionised Western thought. Gravity exists and bodies attract, but the movements of the bodies are relative to each other. Without going into detail, one of the ideas that relativity establishes, although not clearly, is that objective reality is questionable.

By 1918, Planck received the Nobel Prize in Physics. A few years later, in 1922, Heisenberg, Dirac and Schrödinger proposed a new theory, quantum mechanics, in which particles do not have defined positions or velocities. They are assigned something called a wave function. With the wave function there is uncertainty, but you can choose either greater certainty of position with less certainty of velocity, or vice versa. That is, it is possible to know the position and velocity of the particle with a certain degree of accuracy, but there is what is called the "uncertainty principle". Niels Bohr, Wolfgang Pauli and Albert Einstein are extremely interested in the subject and conduct studies based on quantum mechanics.

A funny thing about quantum mechanics—in Schrödinger's opinion—is that all its discoveries bring it closer to Eastern thought than to Western thought, although the latter, stable and objective, is the one that gives science its foundation.

By then, as we have already seen, physics had been divided into two areas of importance: that related to Einstein's theory of relativity, and quantum mechanics.

The two theories—the most advanced in modern physics—have come completely independently, then, to the conclusion that the principle of objective reality makes scientific investigation difficult or impossible.

It is possible, then, to conclude that the human individual at some point finds it harder to study nature exclusively on the basis of his observation, as if he were outside it.

A little more about quantum mechanics. Recently—at the beginning of the millennium—a series of alternatives to objective reality began to be given in the interpretation of this discipline. They are collectively known as Quantum Bayesianism. Among them is what is known in English as QBism (or cubism), which is a form of anti-realism. Its adherents call it participatory realism. That is, the results of a measurement are personal experiences of the individual who measures. Of course, among physicists there is a rather negative reaction to the cubist approach. Something similar is the approach supported by Rovelli, which is called Relational Quantum Mechanics, in which different observers can give different measurements of the same state and all can be approximately accurate.

Cubism and relational quantum mechanics—the result of years of quantum frustration, seem to have finally discovered the fundamental principle of relativity, intuited by Einstein and understood by Schrödinger: the perspective of the individual distorts. It's not that quantum superpositions and particle entanglement are confusing in themselves. What confuses us is our perception based on objective reality, Newtonian mechanics and anthropocentrism, product of Christian and Western philosophy.

There have been numerous attempts to prove the existence of God from the infinity of prime numbers. What they say is that, if the amount of prime numbers is infinite, only an infinite mind can have created them. But the absurdity of the concept is that numbers are symbols that humans have created to understand the idea of quantity. Without human consciousness there would be no numbers, no mathematics, no quantity and, as we have seen, no time.

Søren Kierkegaard said that, by coexisting, the universe and the human mind cause absurdity. He wasn't so wrong. Our perception of our surroundings is extremely limited. As limited as our senses.

Finally, there have been—as we shall see—studies on the physical location of consciousness in the brain and on the historical origin of consciousness, that is, we have wanted to understand consciousness by placing it in space. Consciousness and time, however, do not seem to have a physical origin that can be studied by observing them according to the Western paradigm.

Either way, Rovelli says reality can't be understood by observing objects alone. Totally agree.

The latest confirmation that science is moving away from objective reality is the 2022 Nobel Prize in Physics. Aspect, Clauser and Zeilinger shared it by demonstrating entangled quantum states, that is, they proved that, given certain parameters, photons totally separated from each other behave as a unit. What these scientists were able to demonstrate is based on a work by John Stewart Bell, from 1964, itself based on the work of Einstein. Definitely, the denial of objective reality—even if it was not recognised at the time—was part of the theory of relativity during the early twentieth century.

We have seen a bit of quantum mechanics.

To round this up, let's see if we can summarise the evolution of quantum mechanics in a few intelligible paragraphs and if we discover something interesting about time and/or consciousness. I'm going to try to make it as light as possible. I believe that, if I understand some of it—and I understand very little—that bit can be understood by anyone, even if they are not scientists.

Well, we have described, I think, in a rather intelligible, logical manner, the transition from Newtonian mechanics to quantum mechanics—the latter, which fails to coincide with the stable rules of science or with objective reality. The search of physicists today, as we have already seen, is to find a theory of what, at one time, was called "the unified fields", which is now called "quantum gravity", an attempt to find the logical union of the theory of relativity with quantum mechanics. That is, to ensure that the reality of the universe at the classical level or astrophysics—immense—and reality at the infinitesimal level can be described using the same principles and the same rules. Something holistic.

One of the ways science works is the discovery of certain regularities in nature's behaviour. Based on these regularities, laws are established, which are sometimes maintained, sometimes become obsolete, and sometimes change gradually as new discoveries are made. At first, the regularities that were the easiest to discover were those that had to do with the movements of celestial bodies. Then, science came to the study of the smaller things, until it reached particles.

To measure a particle, you have to put it under light. But it cannot be any amount of light or any kind of light. It has to be, at a minimum, a "quantum"—a measure of light—and it has to be short-wave light. That light has higher energy than visible light, which affects the speed of the particle.

Werner Heisenberg then determined, in 1927, that the position and velocity of a particle could not be measured simultaneously and that, if that happened, it could not be done accurately. That's called Heisenberg's Uncertainty Principle.

There were several reactions from scientists to the uncertainty principle, some of whom found it difficult to accept the modification of the physical theory; Einstein was among them. At that time, Heisenberg, Dirac and Schrödinger proposed a new theory, that of quantum mechanics, in which particles do not have defined positions or velocities. They are assigned something called a "wave function." With the wave function there is still uncertainty, but you can choose either greater certainty of position with less certainty of velocity, or vice versa. That is, it is possible to know the position and velocity of the particle with a certain degree of accuracy, but the validity of the uncertainty principle remains.

Quantum mechanics (in this case, Rovelli) tells us—for example—that time is granular, i.e., that it is composed of discrete elements, although it does not specify what these elements are called (moments? instants?), whereas until now we had always thought, like Heraclitus, that time flowed. But then, we have already explained the main gist of our thoughts on time.

Quantum mechanics notes the intrinsically relational nature of the universe. Everything is related; everything is connected.

Based on relativity, it seems that we should agree with it. Individual human consciousness, for example, is definitely related to collective consciousness, in space and in time.

Our thinking and behaviour are influenced by both contemporaries and predecessors. At the same time, we can influence our contemporaries and our successors. As we have seen, however, science has not yet succeeded in totally establishing the dual nature of consciousness.

Quantum mechanics notes that the observer exerts influence on the observed and vice versa, resulting in a complication that science and technology seem to be unwilling to accept. The tree that falls into the forest without witnesses makes no noise because noise requires a listener. Without someone to hear, noise does not exist. The same goes for all the senses. Without a conscious being, nothing exists the way we perceive it. In reality, everything ceases to exist, especially the past and the future. Especially time.

At one point, when discussing Heisenberg's thought, Einstein notably says *"God does not play dice"*, and suggests a theory of hidden variables, in which the existence of unknown parameters would be accepted.

In *Brief Answers to the Big Questions*, Hawking rejects hidden variable theories and, ironically, also quotes St Paul, when he says that humans only see reality *"through a dark glass"* (a reference to ancient mirrors, in which reflections were dark and distorted). Unlike Einstein—who seemed to believe in a vague God and defined himself as "non-atheist"—Hawking does not believe in God and, although he speaks of "a dark glass," he seems to believe that human knowledge has no limits.

That's another thing about quantum mechanics: it has strange contradictions—it's part of the scientific realm and doesn't seem to accept objective reality, at least not entirely.

On the other hand, the Theory of Relativity does not seem to accept objective reality either, although the scientific community had to accept it as something undeniable.

These scientists assert certain things that, most of the time, would seem impossible to prove. For example, they say *"information is not lost."* It's not clear what kind of information they're talking about. It's physical information, of a binary type, I imagine.

ERWIN SCHRÖDINGER, one of the most important scientists of the twentieth century, won the Nobel Prize in 1933. He was a physicist, but he was fascinated by Indian philosophy and believed that, in the universe, mind precedes matter. Just as Wallace was in the nineteenth century, he was anti-materialist.

Schrödinger, was, rather than a just scientist, a multifaceted genius. He became interested in genetics, the phenomenon of life, ethics, religion, and the philosophical aspects of science. He became interested in the problem of consciousness. He said: *"Consciousness cannot be considered in physical terms. Because consciousness is absolutely fundamental. It cannot be considered in terms of anything else."* He claimed to base his idea on the fact that *"consciousness is never experienced in the plural, only in the singular. Not only has no one experienced more than one consciousness, but there is not the slightest circumstantial evidence that it has happened anywhere in the world."* He also said: *"Obviously there is only one alternative, namely, the unification of all minds or consciousnesses. Its multiplicity is apparent, the truth is that there is only one mind.*

That's what the Upanishads say." I thought that we are all aspects of a single mind that forms the essence of reality. I totally agree with him. He referred to this as the *"doctrine of identity."*

In addition to physics, as we said, he was interested in things like the problem of mind-body dichotomy, free will, and objective reality.

Schrödinger's breakthrough in the field of genetics was something of extreme importance. Until then, biologists spoke of the gene as a hereditary unit without further definition. Today it is understood that genes have a code that programs the structures of cells and establishes what happens to living organisms. When he formulated his hypothesis, Schrödinger used a different approach because he was not a biologist. Schrödinger used, instead, his knowledge of quantum physics.

Schrödinger came to extraordinary conclusions. He believed—like Borges—that a man is all men. Why could one believe oneself differently from others? What makes us so unique, apart from experience?

But I find it impossible to keep talking about quantum mechanics and time without mentioning—again—Carlo Rovelli. I have a lot to say about his book *L'ordine del tempo*. It is a must for anyone interested in time. His description of quantum mechanics' encounter with time is spectacular. I am fascinated by the chapter called *The Aroma of the Madeleine*, in which he talks about Buddhism, about St. Augustine and the beauty of his work, about Proust and about suffering. *"Time is suffering,"* he says. In that moment, Rovelli takes quantum mechanics towards beauty, feeling, meaning. Science makes sense. The universe makes sense.

So far, in nearly a century, the search for the "unified fields" seems to have led nowhere; "quantum gravity" looks like another Holy Grail. Physics has run into the wall of the unknowable and the ineffable. *"Philosophers continue to discuss the disappearance of the present,"* Rovelli tells us. *"The whole idea that the universe now exists in a certain configuration and changes with the passage of time is no longer useful."* But there is no answer.

Newton believed in a time that flows always in the same fashion, uniformly, regardless of change. Leibniz rejected the concept, and followed Aristotle: time is the measure of change. Einstein, with genius, discovered that time (only perceived within high consciousness) and space (only perceived through the senses) are one and the same, and that they are relative. The gravitational field is also relative. As time is. And it exists.

Rovelli gets to the "quanta". With them the granularity, indeterminacy and relative aspect of the physical variables of time are discovered. That is, that time does not flow continuously, that there are many times that overlap, and that it materialises at certain moments with respect to certain objects. The matter gets complicated, and more than clarifying anything about time, it makes it more confusing.

Based on that, Rovelli comments: *"None of the pieces that time has lost (singularity, direction, independence, the present, continuity) call into question the fact that the world is a network of events. On the one hand, there was time, with its many determinations; on the other, the simple fact that nothing is: things happen."* The evolution of science, then, suggests to us that when we think about reality, we must think about change, not permanence. Not in being, but in becoming. Nothing is still. Everything moves.

And if time is becoming, everything is time. There is only what exists in time. But then Rovelli seems to want to clarify that one cannot think of a succession of presents. The most we can say about reality is that the present is relative to a moving observer. Here, definitely, like relativity, he denies the existence of an objective reality. We are getting closer and closer to Eastern philosophy.

What seems contradictory is that Rovelli, after saying that time at the quantum level does not exist, tells us that *"the distinction between past, present and future is not an illusion."* He then explains that the distinction is because the grammar of the languages we use is inadequate.

Well, the languages we use are not inadequate. They describe reality as we need them to describe it. There is a distinction between tenses. We need to have that distinction to live in society, to be able to explain that events happen and when they happen.

We don't need to have a variable called "time." Things, Rovelli tells us, are now much clearer. Things do not evolve over time, but change relative to each other. Why not talk about 'change' rather than 'time' then?

According to Rovelli, there is thermal and quantum time. That variable is what we call time, he tells us, even though there is no variable at the fundamental level.

He adds: *"The time of physics is, finally, the expression of our ignorance of the world. Time is ignorance."* Totally agree.

I want to close this chapter with one last quote from Rovelli: *"This, which is a fact, opens the possibility that it was not the universe that was in a certain configuration in the past. Perhaps it was us, and our interactions with the universe, that are determined. We are the ones who establish a certain macroscopic description. The initial low entropy of the universe, and therefore the arrow of time, may have more to do with us than with the universe itself. This is the basic idea."*

Ah, then, we are the ones who determine what is order and what is disorder. Order is a human concept. Totally agree. So, the arrow of time has to do with consciousness. But we'll talk about that later.

CHAPTER 4

GREEK PHILOSOPHERS AND THE BIBLE
(ANOTHER OPTIONAL CHAPTER)

> *No man ever steps in the same river twice,*
> *for it's not the same river and he's not the same man.*
> - Heraclitus of Ephesus

Before Greece's trio of leading thinkers —Socrates, Plato, and Aristotle—there were those we now call the "pre-Socratics." There were a few: Xenophantes, Heraclitus, Pythagoras, Parmenides, Democritus, and Thales of Miletus, among others. What they did was to turn away from mythical, religious thought, and question it. They were all deep observers. Everyone was interested in the origin and nature of things. To a greater or lesser extent, everyone was interested in consciousness and time although, in reality, until Aristotle, no one really discussed time, except tangentially. Aristotle thought that time only served to measure change. Although one may not agree one hundred percent, let's say that everything changes and that you have to be able to express and measure it in some way.

Among the best-known pre-Socratics, of course, is Heraclitus, who had an incredible capacity for observation. One of the things he noticed about time and consciousness, and that caught his attention the most, was that "everything flows, nothing remains." He first noticed the instability of things, the constant change of nature. Then, he came up with the famous river aphorism. Consciousness and time.

Again, that's something I thought about many times when crossing the Riachuelo between the southern suburbs of Buenos Aires and my then place of work. The river did not seem to flow much, and during adolescence one has the idea that nothing really changes but, after the age of twenty, I began to realise that Heraclitus was not so far off the mark.

Greek philosophers talked a lot about time and its flow. Plato thought that time had a beginning and an end, an idea that Christianity adopted it without hesitation, as we will see when we discuss the Bible. The concept of a linear time, with a beginning and end, fit well with the idea of the Last Judgment and was useful to give a reason to life. But that came later.

For their part, Parmenides and Zeno thought that time was an illusion. Some, like Antiphon, believed that time did not exist. A bit like Aristotle, Antiphon thought that it only existed to measure the world. They were both right in a way.

We have already seen, very briefly, Aristotle's thinking regarding time. And we will see that, although he says that time is not change, he defines it, basically, as the measure of change.

The Greek philosopher—who was born in the province of Macedonia—proposes some time-related dilemmas. Time cannot exist, he says, since it is divided into two: the past, which is no longer here, and the future, which has not yet arrived. He also says that the "now", the present, cannot always be the same because it always changes, it is always different. Here, we do not agree one hundred percent.

But there is time only when there is change. Without change, time does not exist. It is something essentially related to time. The philosopher refers to the causality of time. If there is change, there is time. And if there is magnitude, there is change.

There is an "order of time," as Carlo Rovelli says. Although Rovelli does not seem to entirely agree with Aristotle, because the Greek speaks of time not being a discrete plurality (that is, it does not consist of many entities). In saying this, he says that time is a continuum, while Rovelli, as a good quantum physicist, says that time is granular.

The Greek says that time measures change, and that it is also measured by change. The somewhat enigmatic definition he gives is that time is a kind of number that measures change. If time occurs only within *psyche*, that makes sense.

There are instances when he seems to contradict himself with respect to the present: he says that the "nows" before and after are different, but that there is also a way by which they are the same thing. The comparison he makes is that, if there is something in motion, at first it is one thing, and when it ends it is still the same thing, but different.

Some things don't reside in time, he says. Only things that last for a while, exist in time. That means that eternal things are not in time. However, he believes that there are things that move eternally (I imagine he would be referring to celestial bodies). And eternal movement is part of time. However, there is one way that eternal things are not in time: time does not make them age or decay. What time does, then, is introduce causality.

But then he makes consciousness appear. Aristotle wonders if there can be time without beings with souls. (Yesss!) Time, he says, depends on the soul while change does not depend on the soul. Time is something that can be counted and that, in reality, is counted. Therefore, to exist, it needs beings who can tell it. Time is a number because it is countable, and it is only countable when it is counted. Here, I humbly agree with Aristotle. Without consciousness there can be no time. And without time, there is no consciousness.

I think that, for this chapter, we have had enough of Greek philosophy.

Let's talk a bit about the Bible.

∽

ORIGINALLY, the ancient Hebrews appear to have had no concept of "soul" (as in "high consciousness") as such. They only had a notion of "life." Let me explain the difference. In the Old Testament, in the Tanakh, or Jewish Bible, the term "*nephesh*" is used numerous times, which means "breath of life", which is what God gives Adam when he creates it. What makes him a living being.

At one point, however, the ancient Hebrews saw that there was a difference between the consciousness of a human being and that of an animal. If you read carefully, you will notice that there are two creations in the Tanakh. In the first one, God creates Adam from dust, as we said, and he gives him the "breath of life". In the second one, God creates Adam and Eve and places them in the Garden of Eden.

The term "*psyche*", for "soul", is a Greek term. By the time the New Testament is written, the word "*psyche*" begins to be used as "soul" (or "consciousness").

A couple of generations after the Hebrews fled Judea for the Diaspora, when they actually settled in the Hellenic territories that include Egypt and present-day Turkey, many of those born in those territories began to forget Hebrew and Aramaic, the languages of their ancestors. The first translations of the Bible into Greek are made during that time.

The mother tongue of St. Paul—who was a Jew of the Diaspora—was Greek, although he also knew Latin, Hebrew and Aramaic. According to St. Paul, Adam and Eve were immortal until they ate the fruit of the Tree of the Knowledge of Good and Evil, that is, until they had consciousness, until they ceased to be animals to become human beings. At least, that's how I interpret what the Book of Genesis says (*The Myth of Adam and Eve and the Endurance of Christianity in the West*, 2021). That is, they were immortal until, with the use of reason, they discovered that time existed, they began to remember the past and imagine the future, and they knew that they were going to die. In the myth, to give an idea of what consciousness is, the Hebrew scribes speak of "Knowledge of Good and Evil."

In the Old Testament, Ecclesiastes talks a lot about time. Perhaps the most beautiful and most famous verses are those that say that *"Everything has its right time; There is a time for all that is done under heaven. Time to be born, and time to die; time to plant and time to harvest what has been planted." (Ecclesiastes 3:1)*

Ecclesiastes, too, explains how God granted temporality to man: *"God made everything beautiful in his time, and put in the human mind the meaning of time, even when man fails to understand the work that God does from beginning to end." (Ecclesiastes 3:11)*. The interesting thing about this verse is that God puts time in the human mind, not making it part of reality but through consciousness. The Hebrews, then, seem to have understood very well the close time/consciousness relationship.

Ecclesiastes says, *"A good way to acquire wisdom is to learn to live each day with an eternal perspective. Our Creator has placed eternity in our hearts (Ecclesiastes 3:11)*. Again, my interpretation is that the Hebrews considered consciousness as something that surpassed the physical realm, that is, consciousness as something universal, shared in time and space, but especially in time. To make it clearer, the relationship that Aristotle's mind may have with ours. The way it has influenced us.

In Psalms, Moses prays, *"Teach us so to count our days, that we bring wisdom to the heart" (Psalm 90:12)*. That is, if we do not consider our actions and the consequences of them over time, we will not have wisdom in our hearts.

The Jews were a suffering people, and the Bible speaks of that. Wars, Assyrian occupation and exile in Babylon. God seemed to have abandoned them. The enemies of the Jews always seemed to win. Someone, then, came up with the idea of the Last Judgment. At some point, at the end of days, when time ran out, the Apocalypse would come and sinners would pay for their sins.

The idea of the Last Judgment is closely linked to the heaven of Christianity. That gives meaning to life. The good go to the Kingdom of Heaven. Moreover, at the end of time, not only are our lives given a purpose (the teleological background of Christianity), but the existence of the universe, the directionality of time, and the conduct of God are explained. In Judaism, as in Christianity, time is neither an illusion nor a cycle. It has a glorious beginning and an end. God is in charge of a plan.

And perhaps talking about the Bible in this book serves, among other things, to connect Christianity with science and time. Creationism, as we will see elsewhere, places the human being as separate from the rest of creation. In the Judeo-Christian tradition, man is not an animal like others. It is the reflection of God. In Christianity, God is also man. And there is one ultimate purpose, which is to reunite with God in the Kingdom of Heaven. There is a narrative. Everything has a reason. The implication of this philosophy is that our lives have a purpose. There is a certain order hidden within the chaos in which we live. That's called "Sacred Time." The Christian view of the time is that we have to be optimistic about the future. All Western culture is based on that point of view. Life is progress.

But enough of the Bible. In the next chapter I speak of St. Augustine, not because of his Christianity, but because he was an important Western thinker and because he was very interested in consciousness and time, as we shall see.

CHAPTER 5

ST. AUGUSTINE

St. Augustine talks a lot about time; he does it with total clarity, and with the clarity of someone who has thought a lot about the subject:

"Perhaps it can be said with correctness that there are three times: a present time of past things; a present time of present things; and a present time of future things. Because those three coexist, in a way, in the soul, since otherwise I cannot see them. The present tense of past things is memory; the present tense of present things is direct experience; the present tense of future things is expectation."

In Book X, before dealing with time, St. Augustine begins by talking about consciousness. He speaks to God: *"... Nor do I do it with the words and sounds of the flesh, but with the words of my soul, and with the lament of thought, which Your ear knows. ... My confession then, O my Lord, which I make in Your presence, I make in silence, and not in silence."* That is, he speaks of the words that exist in his mind, that he can hear, and no one else. When you think about it, it is something extraordinary: the voice of the mind. Because we have ideas and we think in terms of time and images, but we also think with that inner voice that tells us things. He´s discussing high consciousness.

Which reminds me—and now I digress a little, as usual—of a description that St. Augustine makes of St. Ambrose, and that Alberto Manguel also quotes in his book "*A History of Reading*": "*But when he read, his eyes glided through the pages and his heart searched for meaning, but his voice and tongue rested. Often, when we went to see him (because no one was forbidden to enter, nor did he want anyone to be announced to come to see him), we would see him like this, reading to himself...*" At a time when people would only read aloud, St. Augustine marvels at the possibility of doing so in the realm of consciousness.

How come I have consciousness, he says, and animals don't? "*I turn to myself and ask myself, 'Who are you?' and 'One Man' will answer. And I find that in me there is a soul, and a body; one outside, and the other, inside me. ... Animals, large and small, can see the body, but they cannot be asked, because they have no use of reason besides their senses to judge what they see. Men can do it...*" The interesting thing is that Augustine sees himself as "inside" his body, and his soul as "inside" himself. In other words, it places identity in an intermediate place between body and soul.

The association with time is not long in coming. St. Augustine begins to speak of his memory: "*And I go to the fields and to the spacious palaces of my memory, where there are innumerable treasures of images, arising from things of every kind that my senses have perceived."* With his incredible capacity for introspection, St. Augustine speaks of the possibility that memory gives him to travel in time. You can take out of memory all the things you have seen and bring them into the present. Of course, they're just images, he says, not the things themselves. And thanks to memory he can sing silently and can distinguish between the aroma of lilies and violets, even if he is not smelling anything, and he knows that he prefers honey to sweet wine, that he likes soft things and not rough ones, all this without liking or touching anything in reality: only in his memory.

As the reader can see, he marvels at the images and sensations that appear in his mind, the product of memory. Here again, I'm going to digress a bit, because I think the way St. Augustine speaks of consciousness—in itself, as he differentiates ideas, images, and words—deserves a little analysis.

A Google engineer recently said that a robot had transformed into a person. Of course, there was quite a stir. These robots have algorithms that allow them to "understand" words and be able to predict what the interlocutor says in order to converse. The robot in question was so impressive that the engineer began to think it already had consciousness. But language has a limited nature. For a long time, we believed that thought was exclusively based on language.

Language is a very elementary codification of what thought is, despite what was believed in the twentieth century. Neither Siri nor Alexa can think like a human—that's easy to understand. What robots comprehend about our world is something very superficial. At that time, it was believed that, basically, mental images were something totally grounded on language, and language, whatever it may be, does not include all knowledge. There is a big difference between reading a piece of music and hearing it or, even more, interpreting it. If anything was discovered with Chomsky's ideas, it was that language is not analysable as a mathematical formula. There is something underlying thought, in consciousness, that is far from linguistic. Human beings often understand each other in a non-linguistic way. Robots have a superficial understanding of what we say: they only understand language. They have no practical experience of the world and no sense of how it works. If we leave aside the concept that knowledge is just a linguistic phenomenon, we realise everything else that has to do with our thinking. We acquire a deep knowledge by observing the world, exploring it and dealing with other human beings. That is perhaps one of the shortcomings of artificial intelligence (AI), which is given written information, so that AI works only on the basis of language, only information, which is extremely shallow. For AI to reach the level of human consciousness, it will first have to have sensations, and remember images, and feelings (*"the fields and spacious palaces of my memory"*); it cannot do that because it is only cultural. AI cannot perceive. It does not have a body. We can safely say that AI will never be comparable to human consciousness. It lacks one of the layers, the basic one.

Let's go back to St. Augustine, who is amazed that he can talk about things. And sounds may be in Latin or Greek, but things exist neither in Latin nor in Greek, nor in any other language. They exist in themselves. He sees the lines that architects make, which represent things, but those are not the things that his senses see, and yet he can recognise them. That is, human beings can recognise symbols.

In that sense, Augustine thinks that universals, such as numbers, mathematics, and logic exist in the divine intellect which, rather than infinite, is eternal. That's something on which I cannot agree with him. Numbers exist only in human consciousness, as much as the concept of eternal and infinite. To say, for example, that primary numbers are infinite is something we devise with our human consciousness in order to explain both those numbers and the very concept of infinity. The moment human consciousness disappears, the number one, twenty, or five hundred and eighty-four thousand will disappear as well, because they are just explanations that we need to give ourselves in order to understand something. They are ideas. Universals, then, as much as time, and other concepts—I propose—exist only in high consciousness.

Today, many scientists see science as a series of "discoveries" regarding the workings of nature. In reality, the "laws" that science discovers are only very useful explanations of phenomena that are repeated in nature. That's not to say that nature created these "laws." Like numbers, the laws of nature are of great use to us. "Laws" are a result of the relationship that exists between the human mind and nature. They are part of the observer-observed relationship. That is, our utilitarian interpretation of something that happens in nature.

I'll give you a reverse example, a leading neurologist tells us: *"That means that, no matter how complicated a discovery is, shortly after doing so we realise that nature already knew about it and had taken advantage of that phenomenon. Long before Newton discovered the laws of gravity, our muscles were stretched in calculated attempts to defy it, and our eyes followed objects in the sky that went at speeds of about ten meters per second squared. ... Long before Einstein explained relativity, our brains coordinated specific times with reference to the observer."* Really? Sorry, sometimes scientists seem to want to reverse causality. Observing the detail so much, they lose sense of perspective at a higher level. Let's continue with Augustine.

The great Christian philosopher had possibly studied Greek theology and philosophy, maybe the *"Theogony,"* the poem in which Hesiod describes the traditions of the gods, and also Plotinus, whose kind of introspection seems to have influenced him.

St. Augustine then speaks of psychology, of how time passes in his consciousness and of how he can travel in time, towards the past and towards the future, with his memory and with his expectations. He marvels at the way his identity is maintained over time. Time is a mystery, he says, *"My soul burns trying to understand this complex enigma."*

Memory goes towards the place where change can come from, so that it can anticipate. According to science, our consciousness consists of systems of perception that include interoception (that is, awareness of the internal state of the body) and exteroception (the senses that perceive the outside). In the Augustinian concept, the human being is, more than anything, a being with a high degree of self-perception. The human being perceives himself and, in doing so, intuits identity, the essence of the past, with which he can understand the presence of the present and the anticipation of the future.

In Book XI of his "*Confessions*", the philosopher begins to speak of the beginnings of time and of God, and tells us what others ask him: *"What did God do before he made Heaven and Earth? If he hadn't done anything until then (as they say), why did he decide to do anything? Did a new activity and a new desire to make a creature arise in Him that He had not done before? How can there be a true eternity then, if a will arose that had not existed before?"* And he answers them, speaking to God: *"You do not make time precede time, otherwise You would not precede all times. But You precede all pasts, with the sublime presence of eternity, and surpass all futures, for they are future, and when they come, they will pass away; but You are the Same, and Your years never end."*

And he gives an idea of how the mystery of time works: *"But is there anything we talk about with greater familiarity, and knowing what it is, other than time? However, if no one asks me, I know what it is; If I want to explain it to someone who asked me, I can't say what it is."* Time, part of high consciousness, is as difficult to explain as is high consciousness itself.

The amazing thing about St. Augustine is that he continues to be a source of wisdom and continues to influence studies such as the relationship between bodily processes and sensations, especially through introspection. He wants to show us the Christian idea that the human being is unique in Creation. We have an exclusive power which is the awareness that we exist. And the other thing is the principle of identity: the possibility we have to exercise self-reference.

Discussing time, the philosopher offers a great definition of eternity: the non-existence of time: *"But if the present were always kept as present, and never became past, it would truly not be time but eternity."* So, according to St. Augustine, if the present continued indefinitely, it would be transformed into eternity. From a current perspective, then, eternity could be two things: either something fantastic or a nightmare. A bit like *"Groundhog Day"*, the time Bill Murray can't escape from.

THERE ARE current philosophers who propose two versions of time, one called Series A and the other, Series B. The first version proposes that every moment is past, present or future. The second, that each moment is before or after each other moment. Let's say, a relative time. A parallel can be drawn between these two times and what St. Augustine says about human time compared to eternity. In relative time, if X always precedes Y, there is no change, and therefore there is no time. A bit like the unchanging present of St. Augustine. A little also what Aristotle says, that time is change. In that respect, St. Augustine is Aristotelian, of course.

He explains: *"I once heard a wise man say that the motion of the sun, the moon and the stars constituted time, and I disagreed. Why not consider that time is represented by the movements of all bodies? ... God gives men the possibility of seeing in small things what is common to great things and small things."*

And the philosopher speaks again of the non-existence of past and future as entities separate from time: *"If there is past and future, I desire to know where they are... Wherever they are, and whatever they are, they exist only as present."*

To end with St. Augustine, let's say that another important thing he notices about time—and in this he agrees with Schrödinger and quantum physics—speaking to God, he says: *"Let them see, therefore, that there can be no time without someone created, and let them stop talking vanities. Let them also understand the things that have existed before; and may they understand You, who were before all time, the eternal Creator of all times, that no time can be eternal with You, nor can any created being be eternal either, even if You had created someone before You created time."* There is no time without consciousness. There is no observed without an observer, he tells us.

CHAPTER 6

NEWTON AND LEIBNIZ

> *"Gravity is unitary and indifferentiable.*
> *Gravity is integral,*
> *inclusively embracing and permeating;*
> *it cannot be focused or has a shadow,*
> *and is omni-integrating;*
> *These are all characteristics of love.*
> *Love is metaphysical gravity."*
> - Buckminster Fuller

Newton was born in England in the mid-seventeenth century, and at his death, in 1727, the Age of Enlightenment was already beginning its heyday. In general, although he was mainly interested in biblical exegesis and alchemy, we identify him as the Father of Science.

Newton, along with Francis Bacon, and René Descartes, began the philosophical-scientific revolution that preceded the Age of Enlightenment and the amazing development of the West.

During the beginnings of Christianity, the Church persecuted the doctrines it called "heresies." It was what we now know as the Dark Ages. Driven by curiosity and greed, astrology and alchemy began with forbidden observations and experiments. Secretly, they gave birth to astronomy and chemistry; other, more specialised, scientific disciplines grew from them. Science then crept slowly out of a nebula of superstition. In its early days, the West developed strangely. It was like a dance in which religion, philosophy, art, and finally science, acted as the four bases of Western DNA, which intertwined to form the double helix of our culture. Newton would be initiator and reference in this transformation.

Somehow, scientific work interested him less than the rediscovery of ancient wisdom.

According to the economist Keynes, who was a great admirer, *"Newton was not the first in the age of reason, but the last of the magicians."*

Newton based his classical mechanics on the idea that space is something different from the body and that time passes uniformly no matter what happens in the world. That is why he always spoke of "absolute time" and "absolute space". He did this to distinguish them from "relative times" and "relative spaces," which are the ways we measure those entities.

At the beginning of his work "*Principles*", Newton includes a section, entitled *"Scolio"*, in which he describes his ideas about time, space, place and motion. In that section he explains that absolute, true, and mathematical time passes inexorably, without taking into account anything external, that is, without reference to changes or ways of measuring it. From his point of view, something similar happens with absolute space, true, and mathematical, i.e., relative spaces are measures of absolute space.

Aristotle had rejected the possibility of emptiness, saying that emptiness is nothing, and that nothingness cannot exist. Same with time without change. Until the seventeenth century, these were indisputable truths. Later, the new ideas of Copernicus and Galileo, on the movement of celestial bodies, introduced doubts about what was believed until then.

Newton copiously read and annotated the writings of Walter Charleton, and developed his ideas on the basis of many of Charleton's ideas. Among other things, as we said, that time and space are real entities, although they do not fit the traditional categories of other substances; time flows eternally without being disturbed by the acceleration, delay, or suspension of any physical thing; time has nothing to do with any measure; space is absolutely immovable and incorporeal; movement is the migration of a physical thing from one place to another.

Astronomical observations were part of all this. Newton believed that an astronomical equation of time was needed. The Earth's rotation, he says, is uniform, but it might not have been. There might not have been uniform movements to measure time. That is, the movements of things could be accelerated or delayed. In contrast, time remains unchanged; the speed (fast, slow or zero) of any movement does not alter it.

In his "Principles," Newton elaborates on space, properties, causes and effects, and his arguments are sound. Although thinkers such as Leibniz and Berkeley harshly criticised him, his ideas dominated physics from the seventeenth century until the Theory of Relativity.

We said that Newton—like Plato—believed in absolute time. Time, in that sense, is like a vessel; things or events exist within it. Something that contains them but is independent of them. What did Leibniz think?

What would happen if there was no movement of any kind, say, for a week? For example, if birds stopped singing, if trains were left motionless between stations, if the movement of people froze in the middle of what they were doing, if the smoke of a cigarette were left floating in the same way. If that was possible, time would be independent of all change. The same would happen if things could happen faster or slower. For example, if what happens in two months were to fit in a month. But that doesn't happen, nor can it happen. Time is not something independent. Leibniz was sure of that. Time is not independent of change. As we have already seen, Aristotle also believed that time was closely related to change.

We are not going to go into the names of the different philosophical theories concerning time. But we can say that those who do not believe in absolute time—for example, Leibniz and Aristotle—believe in relative time. For them, time is nothing more than the temporal relationships between events.

Those views on time are closely related to views on space and movement. If one believes that time and space are relative, what follows is that motion is also relative. And the same happens if one believes that they are absolute: the movement should be absolute.

Why would anyone believe in the existence of absolute time? Because it appears logical. What Einstein thought about relativity was counter-intuitive. All great discoveries are counter-intuitive. Time is an integral part of our consciousness. St Augustine thought time was an extension of the human soul [high consciousness]. Totally.

Leibniz did not believe in absolute time and space because he said that, if they existed, they would violate the principle of indistinguishability (i.e., of not being able to discern between things that seem identical). The discoveries of quantum mechanics make the applicability of the principle something much discussed today.

Leibniz considered differents possibilities of reality. One way would be that everything is as it is. The other, that events pass a second after how they actually do, but they are almost the same. If there were absolute time, the two descriptions would be part of two absolutely different worlds. But those worlds would be indistinguishable from each other. That would violate the principle that if two things are indistinguishable, they are identical.

Leibniz's reasoning attempts to identify problems with absolute time and space. Simplifying what he meant: the concept of an absolute time and space is not something observable, and as such, it is something that is best not to consider.

Among the many differences Newton and Leibniz had during their lifetimes, one was that it would seem that they invented calculus at the same time—a bit like Darwin and Wallace with the origin of species—but they never agreed on who had while they lived. What happened was that Newton was consecrated as the father of science while Leibniz died in disgrace. Over the years, Leibniz's thought began to be considered as something very important as well. Today, the two are believed to have invented calculus independently.

Like Aristotle, Leibniz did not believe in a time independent of the changes occurring in it. Time, according to Leibniz, is relative to change; quite the opposite of Newton's absolute time. And, if time is relative, so is space.

CHAPTER 7

JORGE LUIS BORGES

> *"Your matter is time, the incessant time. You are every lonely instant."*
> - JLB

To read Borges is, among other things, to observe how a lucid mind faces two important, but more than anything, universal issues: consciousness and time. We know that Borges has other obsessions and that he constantly plays with them. He is obsessed with tigers, mirrors and labyrinths. However, all his writings reflect, albeit tangentially, his two fundamental themes which, I repeat, are consciousness and time. And he does it from the perspective of an individual who is Argentine and Western, but who never finishes accepting those limitations and embraces his humanity, like Hesse and Schrödinger, even venturing into Eastern ideas and experimenting with a mindset that ignores the limits imposed by Aristotle.

Borges denies objective reality. He repeats again and again that one man is all men and that to kill one is to kill humanity. Consciousness is one and we share it in space and time. "In short, immortality is in the memory of others and in the work we leave," he tells us.

In *The Immortal* he creates the character of Joseph Cartaphilus, someone who is immortal and remembers being, at the same time, Marcus Flaminius Rufus, a Roman centurion, and Homer. Cartaphilus is three people who are, in reality, one. He is not eternal, he is immortal. At one point the character tells us: *"Being immortal is trivial; except man, all creatures are, for they are ignorant of their death; what is divine, terrible, incomprehensible, is to know oneself immortal."*

In almost all his stories and stories, from *History of Eternity*, to *Funes the Memorious*, to *Garden of the Forking Paths*, or *The Script of God*, Borges deals with the existence or non-existence of time and all its possibilities.

History of Eternity shows us the religious side of someone who claims to be agnostic: *"The universe requires eternity. Theologians are not unaware that if the Lord's attention were diverted for a single second from my right hand that writes, it would fall into nothingness, as if it were struck by a fire without light. That is why they affirm that the conservation of this world is a perpetual creation and that the verbs 'conserve' and 'create', which are hostile to each other here, are synonymous in Heaven."*

If we want to learn what his ideas about time are, all we have to do is read his *New Refutation of Time*. In that essay,

Borges begins by making us see that the title contradicts what he says in the essay: that the continuity of time is an illusion. Time does not contain a succession. Every moment is eternity, which negates the mere inclusion of the word "New" in the title. Here, as we will see, Borges agrees with Rovelli and the experts in quantum mechanics. In other writings, he often mentions eternity, for example, when he says, *"[Eternity] ... theologians defined it as the simultaneous and lucid possession of all past and future moments, and judged it one of the attributes of God."* Other times he references things like the duration of hell, but those opportunities are almost always an homage to Dante, or perhaps Swedenborg.

Borges' reasoning tells us about his search, mixing the idealism of Berkeley with the ideas of Hume. The latter rejects identity. He says that each man is a collection of perceptions that occur one after another with inconceivable rapidity. Both believe in the existence of time: for Berkeley it is a succession of ideas. Hume says it is a sequence of indivisible moments. Borges mixes the two perspectives, takes sides decisively and, in doing so, proposes something new: *"I have accumulated transcriptions of the apologists of idealism, I have lavished their canonical passages, I have been iterative and explicit, I have censured Schopenhauer (not without ingratitude), so that my reader would enter that unstable mental world. A world of evanescent impressions; a world without matter or spirit, neither objective nor subjective; a world without the ideal architecture of space; a world made of time, of the absolute uniform time of the Principia; an indefatigable labyrinth, a chaos, a dream. To that almost perfect disintegration came David Hume."*

Borges' inimitable erudition brings us back to the idea of the ego, which Eastern philosophy and Buddhism have considered illusory for thousands of years. That, he tells us, only denies the notion of time we imagine knowing. He tells us about Zeno's paradoxes, which oppose pluralism and change, and claim that movement is nothing but an illusion. There are truths that seem to reject what is evident to our senses.

Berkeley, on the other hand, (*Principles of Human Knowledge*), denies the primary qualities—the solidity and extent of things—and absolute space. Borges suggests that once we deny the existence of matter and spirit, and deny space as well, we have no right to retain time as a continuity. Time, then, does not exist outside of the now. But, in that now, time is everything.

Once he admits idealism, Borges goes much further: anticipating quantum mechanics by several decades, he explains the dynamic nature of our identity, telling us "*there is no secret self behind faces, which governs acts and receives impressions; we are only the series of those imaginary acts and those wandering impressions.*" Actually, we become in the present.

The *New Refutation* continues with a disquisition on collective consciousness and the unity of humanity: "*... if there is no plurality, he who annihilated all men would be no more guilty than the primitive and solitary Cain, an orthodox view, nor more universal in destruction, which can be magical.*" There is not a multitude of sorrows, there is only one pain. He who kills one man, kills all. The perception of what is real is collective, without being objective. We all perceive everything.

But, in addition, Borges' heroes can also be anti-heroes. In *The Shape of the Sword*, the protagonist is the traitor, who ends by saying: "*Don't you see that I have the mark of my infamy written on my face? I told you the story this way so that you would hear it to the end. I have denounced the man who protected me. I am Vincent Moon. Now, despise me."* For him, we are all and one.

In his writings, Borges expands on granularity, the same as the one in quantum mechanics: according to Anaxagoras, he tells us, gold consists of gold particles, and according to Josiah Royce, everything present is a succession, and he tells us that our language is not suitable to explain time, timelessness, or eternity.

But to try to describe the *New Refutation...* analysing the essay, is an exercise in futility. You have to stick to the work. Recommended reading. That is what I intend to do. It's a luxury.

Borges, who defines his work as *"the weak artifice of an Argentine lost in metaphysics"*, can only conclude all his speculation with a mixture of truth and poetry: *"And yet, and yet... To deny temporal succession, to deny the self, to deny the astronomical universe, are apparent despairs and secret consolations. Our fate (unlike Swedenborg's hell and the hell of Tibetan mythology) is not frightful because it is unreal; It is frightening because it is irreversible and made of iron. Time is the substance of which I am made. Time is a river that snatches from me, but I am the river; it is a tiger that destroys me, but I am the tiger; it is a fire that consumes me, but I am the fire. The 'world, unfortunately, is real; I, unfortunately, am Borges."*

CHAPTER 8

EINSTEIN AND SCHRÖDINGER

The way we perceive space is through vision, through the eyes. The way we measure it is also visual. What happens with time is something totally different. The perception we have of time is almost exclusively mental, that is, it happens in high consciousness. We know that it exists, at least at our level, because it is part of memory—the past—and part of expectations—the future. Of course, we can see how things change, how a creature grows, how a tree grows, how a flower withers or how we age, but those changes are almost imperceptible to the senses. Temporality resides mostly within us, in the mind.

The genius of Albert Einstein was perhaps to be able to glimpse, counter-intuitively, that space and time, rather than two different entities, can be considered the same thing. One of them, space, is physical, tangible and measurable with some precision. The other, time, is abstract, intangible, and only pragmatically measurable, since not even the most advanced atomic clocks can measure it accurately.

Einstein introduced the theory of relativity in 1916, more than a century ago. The idea was extraordinary, so extraordinary that it turned Newtonian science upside down and forced it to change many of its principles.

Einstein realised that time is not just the time we live in, but goes far beyond clocks and calendars. What he did, as we saw earlier, was to dictate that the two, space and time, constitute an indivisible whole. His achievement consisted not in thinking like a physicist, but in thinking like a thinker.

Important innovations usually include considering ideas outside pre-established frameworks or canons, thinking divergently. We already saw how Newton dedicated himself to alchemy, like many others in his time, but he came up with something that no one had thought of until then: he began to apply what we now know as a scientific discipline to his discoveries.

As we said, it is not possible for us to see space-time because the element of time is imperceptible to our senses. It is a cultural element, and what is called the fourth dimension. One of the effects of the fourth dimension is that our speed varies depending on where we are; another effect is gravity.

We understand one-dimensional things, like a point, and two-dimensional things like a square. Three-dimensional things are the most familiar to us: a table, a car, a telephone.

The fourth dimension is something else. It becomes much more difficult for us. Although when we want to meet someone, we give him the coordinates of space: "See you at the corner of First and O'Connell", and add the time: "at four". We do it because that is the reason why we invented time. We need it to function in society. In a civilisation without time everything would be chaos. No schedules, no programs, no flight departure or arrival times. When societies were smaller and less sophisticated, time was vaguer and less sophisticated. Nowadays, for a society to function properly, it needs precise time-keeping.

The theory of relativity works in four dimensions. As we said, Einstein's genius was also to realise that it is something we use in everyday life.

Monisha Ravisetti, in her article *Enter the General Relativity Rabbit Hole: unravelling Einstein's theory that deconstructs space and time*, offers a very interesting example: *"If we put a ball on a trampoline, the ball falls towards the middle and causes the elastic fabric to sink. And if we put a ball on the curved fabric, the ball rolls down and is stuck next to the ball. The trampoline is the fabric of space-time, the ball is the Earth, and the ball is you"*

Everything that has mass causes space-time to be deformed, distorted. That's what Newton meant by "gravity." Einstein looked much further and realised that gravitational distortion applies to all bodies that are in space according to their mass. We now know that space is completely distorted by the number of objects that exist in it. Relativity means that space-time causes objects to move and the mass of those objects causes space-time to curve. That's part of relativity.

This is where things get complicated because, if time is a constituent part of space-time, time also curves, dilates. The uniform and absolute time of Newton and the ancients was an illusion.

According to Einstein, time curves with space. But there is one exception: the speed of light remains constant.

So, when light passes through a curve of space, it is crossing a longer distance than if it were space without a curve. At the same speed. So, if space changes and the speed of light doesn't, what changes? Of course, time. If the speed of light is constant, time has to be the only element that is altered. That is, it moves more slowly, and its slowness is relative to the strength of the gravitational field. The theory of relativity establishes the existence of the fourth dimension, which makes the linearity of time an illusion of high consciousness.

But here we enter the philosophical realm and, as we shall see, there are many other things that can be said regarding the links between time and consciousness, in view of which, I find this whole thing very difficult to understand.

Science bases many of its advances on experimentation, but also, as in philosophy, some of them have their origin in pure speculation. In his youth, Einstein imagined what it would be like to run on a beam of light. The idea led him to the discovery of special relativity. Later, he thought of a man falling, and realised that, when he falls into free fall, he does not feel his own weight. From that idea, he concluded that there was no distinction between acceleration and gravitational attraction. That discovery, which led him to the theory of relativity, is known as the "equivalence principle."

Whatever science says, Westerners (those of us who are not scientists, at least) think daily in terms of linear time. There is a before and after. And in the middle is the present, the today, at this moment. There's something teleological about all that. The influence of Christianity—and perhaps Marxism as well—on the West leads us to see the world as something that has to be constantly improved. Christ and His disciples constantly spoke of the end of the world. The final judgment was very close. And that was going to remedy all injustices. Christianity began as a sect with a definitely eschatological vision.

And yet, if we think about it more closely—apart from the ideologies upon which our civilisation is based, including production, consumption, and the free market, which point toward purpose, toward constant improvement, and toward an end—life is always cyclical, a bit like it was in the ancient and medieval ages. The sun rises every day and sets, also every day; the seasons follow each other; the year ends in December and begins, again, in January. There is a cycle that repeats itself. And, if we look at other sources, Eastern cultures speak of a wheel of time, something that the Stoics also talked about.

Einstein, along with many physicists of his time, spent much of his time searching for "unified field theories," theories for the union of gravity and electromagnetism, but he did so without paying much attention to quantum mechanics.

Herman Weyl and Arthur Eddington had tried various theories of unified fields. Einstein sided with Eddington and, in 1923, wrote several essays on the latter's ideas, especially one in which the fundamental field was not the space-time metric but a twist-free connection. By 1925 he had realised that Eddington was wrong. But, of course, he kept trying to find the solution to the problem until his death in 1955.

SCHRÖDINGER, who became close friends with Einstein, also tried to fix it. Einstein believed that quantum mechanics was far from solving anything. Schrödinger agreed with him. They exchanged copious correspondence on the subject. In 1935 Schrödinger had the famous idea of the cat in a box.

The concept is simple. While the cat is in the box and we cannot see it, hear it, or perceive it, it may or may not exist, which is the same as saying that it exists and it does not exist simultaneously. There is uncertainty as to its existence.

The friendship with Einstein suffered when Schrödinger had an idea, which he published without consulting Einstein, that a rotating mass would generate a magnetic field. Einstein told him that didn't differ much from his theory. After that they stopped writing for three years.

Schrödinger thought a lot about his idea that a spinning mass produces a magnetic field. In 1926 he elaborated the mathematical formula of the wave function—the way in which the position of a wave can be described as a range of positions. Based on that, Heisenberg proposed the Uncertainty Principle regarding the position of particles, and then, Neils Bohr presented an idea that combined much of what physics had seen up to that point: the position of a particle can be described as a wave, and the wave is actually the probability of a position. By marrying that idea with the Uncertainty Principle, Bohr concluded that the properties of particles are totally random. Uncertainty is fundamental in the universe. We have already seen that Einstein opposed this idea, saying that *"God does not play dice with the universe."* Bohr replied, *"Don't tell God what to do."* What has been proven so far is that Bohr was right.

I'm not going to go much deeper into what happened with the development of physics and quantum mechanics, because we've already talked about that. Despite Einstein's opposition, Schrödinger's concept of the wave function remained valid (although, as I mentioned before, the concept of wave function is now being questioned). The particles, which began as simple numbers, when described with equations based on the wave function, ended up as something of infinite dimensions.

Schrödinger, on the other hand, ventured into biology, a science in which he collaborated in an important way with his discoveries regarding DNA. Until his death he continued to advocate greater collaboration between physics, biology, and chemistry, especially to explain the emergence of life from inanimate matter. In doing so, he proposed that living matter was ruled by aperiodic crystals, i.e., that it had a non-repetitive molecular structure. His were the first descriptions of DNA.

Incidentally, a team of Japanese scientists has recently proposed a new theory concerning the emergence of life. The scientists developed an RNA molecule that replicates, diversifies and becomes more complex on its own. The molecule has the capacity to evolve.

As we have already seen, towards the end of his life—perhaps based on his discoveries in quantum mechanics—Schrödinger became closer and closer to Eastern thought. There is not something that exists and something that perceives, he said. The subject and the object are only one. Science has serious difficulties in the study of consciousness because our ego, which feels, perceives and thinks is part of the object it wishes to study. That would seem to create a paradox: there are so many egos, but the world is only one. In reality, individual consciousnesses seem to overlap and create the illusion of the concept-world. The place where consciousnesses overlap is what we consider "objective reality." But consciences are, in reality, one. As Schrödinger explains the "doctrine of identity," each mind is identical to the totality of minds. Individual consciousness—at least that's how I interpret it—is compared to universal consciousness just as a drop of water is to the sea. All minds are one. All times are one: now. There is no before or after. Only memories and expectations in the present.

CHAPTER 9

STEPHEN HAWKING AND CARLO ROVELLI

This chapter was going to be only about Stephen Hawking. I then decided to include Carlo Rovelli in it. Both physicists, Hawking and Rovelli have written brilliant books on the subject of time.

I must say, however, that I had to read Hawking's book twice. It was hard reading. I read Rovelli's book three times in English and twice in Italian. Not because it was hard to understand, but because I found it fascinating. I don't quite agree with it, but it is excellent reading.

In his posthumous book *Brief Answers to the Big Questions*, Stephen Hawking tells us that, according to quantum mechanics, all space is filled with pairs of virtual particles made of matter and antimatter. Those particles attract and reject, they are entangled. Although they cannot be seen, their existence is confirmed by the so-called Lamb Effect, something they produce in the energy spectrum of certain light. Now, if there is a black hole and one of those particles falls into that black hole, the other particle (either matter or antimatter) is abandoned. It can fall into the black hole as well, or it can escape to infinity, where it appears to be radiation emitted by the black hole. It can be seen in another way: let's say that if the antiparticle is the one that falls into the black hole, that can be taken as a particle that leaves the black hole and travels backwards in time. When it reaches the point where the particle/antiparticle pair has originally materialised, it scatters through the gravitational field, so it travels forward in time.

Apart from the Lamb Effect, believing that all this happens because it can be tested with physical formulas is very similar to believing that the Son is not created, but eternally begotten by the Father and that the Holy Spirit is not created, nor begotten, but proceeds eternally from the Father and the Son (as claimed by the Catholic Church) or from the Father alone (as the Orthodox believe). That was what the Council of Nicaea agreed to, or failed to agree to.

The purpose of the paragraph above is not to attempt to discredit Stephen Hawking—whose high contribution to science is known to all—but to highlight the irrelevance of certain discoveries of quantum physics. At what point will it be possible to test what happens to the relationship between the particles that fall into a black hole and the particles that are entangled with them? If all the other predictions of the formulas can be made and fulfilled, what relevance could this discovery have?

Hawking was very interested in time, and to some extent in consciousness. In *Brief Answers...* He tells us about the origin of consciousness: *"DNA passes models of life from one generation to another. Even more complex life forms obtain information through sensors such as eyes or ears, and process the information in brains or other systems to figure out how to act, and then act in the world, passing the information to muscles, for example. At some point, during the 13.8 billion years of the cosmos' history, something beautiful happened. Information processing became so intelligent that life forms became conscious. Our universe is now awake, becoming aware of itself. I consider it a triumph that we, who are only mere stardust, have come to such a detailed understanding of the universe in which we live."* With that beautiful description of human consciousness, Hawking expresses his agreement with Sagan: the universe is aware of itself and consciousness is ours. The chapter in which that paragraph is included, unfortunately, is only about artificial intelligence, so Hawking does not elaborate much more on consciousness. He does not appear to have arrived at the notion of the cultural origin of high consciousness and its transmission among human beings within society.

In the book *A brief history of time,* Hawking gives a detailed description of space-time and explains it as clearly as possible from the point of view of physics. At the conclusion of the book, he proposes perspectives and asks fresh questions. In one he rebels against the protocols of science: *"The random, unpredictable element appears only when we try to interpret the wave in terms of position and velocity of the particles. But perhaps that is our mistake: perhaps there are no posi-*

tions and velocities of the particles, but only waves. What happens is that we try to adapt the waves to our preconceived ideas of positions and velocities. The resulting discordance is the cause of the apparent unpredictability. In fact, we have redefined the task of science to be the discovery of laws that allow us to predict events up to the limits set by the uncertainty principle. However, the question remains unresolved: How or why were the laws and the initial state of the universe chosen?"

Hawking questions the methods and goals of science in an even more novel way: *"The normal approach of science to construct a mathematical model cannot answer the questions concerning why there must be a universe for the model to describe it. Why does the universe take all the trouble to exist? Is the unified theory so convincing that it creates its own existence? Or does it need a creator, and if so, does he have any other effect on the universe? And who created him?"*.

Science, he tells us, asks *what* the universe is but, so far, it has not asked *why*. Philosophy, which asks *why*, has not advanced as much as science. Before, philosophy embraced all human knowledge. Now it has been divided and science is too technical and mathematical. It gives us a glimpse that human knowledge must be expanded.

If we discover a unified theory, he adds, it should be available to everyone, not just scientists. Then we will be able to ask ourselves the question of why we exist. If we find the answer, we will have known the mind of God.

One may or may not agree with Hawking regarding his vision of consciousness and time, but there is one thing that is clear about his thinking: to understand these concepts we will need multidisciplinary participation.

Of course, in 2023 the "hard problem" of consciousness remains unsolved. Today I read that, in mid-November 2022, there was a meeting of neurologists in San Diego, California, with more than 24,000 participants. The meeting, the article says, was a tribute to reductionism—solving difficult problems by dividing them into knowable entities —that is, scientific analysis. We continue to use the same methods and protocols to reach a new conclusion. It would seem that we are not quite on the right track.

Now, let's try and see some of Rovelli's thinking. What did I find so fascinating about his book? I agreed with him many times. He says: *"The idea that a well-defined now exists throughout the universe is an illusion, an illegitimate extrapolation of our own experience"*... *"If the present has no meaning, then what exists in the universe?"*. One hundred percent. Much of how we understand reality is that *"illegitimate extrapolation"* that he mentions. We think in terms of time, but time exists only in our thoughts.

Without totally acknowledging that time is a human device that we use to live in society, Rovelli agrees with Aristotle: *"If nothing changes, time does not pass—because time is our way of situating ourselves in relation to the changing of things: the placing of ourselves in relation to the counting of days"*. It's the idea behind *"Groundhog day"*. The only difference is that the movie has an incredible number of readings.

But then, he says: *"Before Newton, time for humanity was the way of counting how things changed. Before him, no one had thought it possible that a time independent of things could exist.*

Don't take your intuitions and ideas to be natural: they are often the products of the ideas of audacious thinkers who came before us." The last sentence is confusing, because he appears to agree with Newton's absolute time. He then clarifies *"But Newton is wrong in assuming that this time is independent from things—and that it passes regularly, imperturbably, separately from everything else"*.

Sometimes, Rovelli seems to understand time clearly, it's just that he doesn't express reality in its entirety, in that time is only a human device: *"There is, nevertheless, an aspect of time that has survived the demolition inflicted on it by nineteenth- and twentieth-century physics. Divested of the trappings with which Newtonian theory had draped it, and to which we had become so accustomed, it now shines out with greater clarity: the world is nothing but change"*. The world is change indeed. We are not, we only become.

And then: *"... our grammar is organized around an absolute distinction — past/present/future— that is only partially apt, here in our immediate vicinity. The structure of reality is not the one that this grammar presupposes. We say that an event is, or has been, or will be. We do not have a grammar adapted to say that an event has been in relation to me but is in relation to you"*. Here it becomes quite clear that Rovelli doesn't quite grasp the concept that time and language are both elements of human *psyche*. Both, language and time are our way of dealing with change within society. Language does not need to be relative to express relativity.

Then we realise that he is talking about quantum mechanics and only about quantum mechanics: *"There is no need in any of this to choose a privileged variable and call it time. What we need, if we want to do science, is a theory that tells us how the variables change with respect to each other. That is to say, how one changes when others change. The fundamental theory of the world must be constructed in this way; it does not need a time variable: it needs to tell us only how the things that we see in the world vary with respect to each other. That is to say, what the relations may be between these variables."* He discusses time in terms of scientific need. But time has nothing to do with scientific need. Time is a device human beings have invented in order to understand reality, plain and simple. Its non-existence is related to life, not science.

In the twelfth chapter of his book, *The scent of the madeleine*, Rovelli hints at philosophy and appears to understand time as part of a social phenomenon: *"We have shaped an idea of a human_being by interacting with others like ourselves. I believe that our notion of self stems from this, not from introspection. When we think of ourselves as persons, I believe we are applying to ourselves the mental circuits that we have developed to engage with our companions"*. Now we're talking.

And there are more hints: *"It is memory that solders together the processes, scattered across time, of which we are made. In this sense we exist in time. It is for this reason that I am the same person today as I was yesterday. To understand ourselves means to reflect on time. But to understand time we need to reflect on ourselves"*.

Rovelli goes on to explain how we can listen to music in a present that is not a mathematical one, but a gooey, a sticky, gelatinous reality. We live in that present and thanks to that present we can enjoy the glory of Beethoven.

To finish this chapter, I want to repeat that, even if I do not agree entirely with Rovelli, I believe *The Order of Time* is a most readable and enjoyable book.

CHAPTER 10

CONCLUSION

"If time is a mental process, how can it be shared by thousands of men, or even two different men?" ... "None of the various eternities that men planned—that of nominalism, that of Ireneaeus, that of Plato—is a mechanical aggregation of the past, present, and future. It is a simpler and more magical thing: it is the simultaneity of those times. The past is in its present, as is the future. Nothing happens in that world, in which all things persist, still in the happiness of their condition"
- Jorge Luis Borges

Consciousness and time have several qualities that science —especially physics—would seem unwilling or unable to deal with. Is it because they are non-physical? Perhaps it is impossible for science to deal with them. Perhaps, in any case, physics tends to impoverish these qualities of consciousness and time in a systematic way because it does not have the elements to describe reality in all its complexity. It falls short of the nature of both, consciousness and time. Among these qualities, I refer more than anything to the intimate relationship that exists between them. The richness of that relationship is lost in a discipline that idealises, that needs to take the flesh out of physical reality and turn it into symbols to understand it better. Philosophy, psychology, and other humanistic disciplines do not need to do that.

We can say with complete certainty—I repeat—that without high consciousness there is no time, and, without time there is no high consciousness, no identity. Science has dealt with the consciousness-time relationship, but only marginally. Because before you do it, you need an exact definition of both—of time and of consciousness. That is impossible in scientific terms since objective reality is a creation of science, as we saw. And the perception of time is, by its nature, exclusively subjective, or subjectively shared. The perception of time—as we have already seen—resides only in individual high consciousness, and is shared from different points in space.

It would appear that when I say that science cannot make greater advances without changing its paradigm, what I'm saying could be compared to the famous saying of Charles Duell, Commissioner of the United States Patent Office in 1889: *"Everything that could be invented has already been invented."* On the contrary, I think there is a lot to discover, it's only that, to do it, we have to think that observation need not be solely analytical. I give an example: in the second half of the twentieth century, Noam Chomsky presented his perspective of a generative grammar that claimed to describe language based on an oversimplification of the syntax of a basic sentence. The basic sentence was then transformed to generate other possible sentences. Transformational generative grammar was fashionable for a short time and then failed. And it failed because such analytical observation leads to dead ends. Dissection is only a first part in the study of physiology. Without dynamic details there is no breakthrough.

Einstein combined time with space by conceiving the existence of space-time. He did it in the field of physics and only as part of a scientific analysis. But he went much further: he said that there is not a single time, but an infinity of times. As many times as there are points in the universe. And that has consequences in our daily lives. That the clock on the table measures a different time than the clock that is on the floor really does not matter at our daily level. That there are infinite times does, especially if we want to define the nature of time to reach some conclusion, scientific or philosophical. That is to say that, by existing in consciousness, time has expressions as varied as there are individual consciousnesses and as the points in space in which those consciousnesses are, or those that can be imagined. Perhaps at some point, a definition will be achieved that includes those many variables and it will make sense.

Another of the coincidences between consciousness and time is the singular and at the same time multiple nature of both. We said that there is one time and infinite times. There would also seem to be a consciousness and what seem like infinite consciousnesses. High consciousness does not exist in the individual in isolation. As a linguist I know that consciousness is shared, as language is shared, which is the same thing that happens with time. That does not mean that there is an objective reality. Reality only seems to exist as the sum of the perceptions of individuals and their interactions with others. And we all have some moment that gives us the circumstance that completes us.

Maria Popova explains the moment that touches, us succinctly but very completely:

"We are living interludes, bookended between not yet and no more, each of us a random draw of the cosmic lottery, each allotted a sliver of spacetime in which to live out our lives as chance configurations of stardust suspended in time."

Scientific analysis admits questioning, but it needs to be exhaustive in terms of imposing limits. Science divides to understand. When things are expected to be a certain way, unfortunately, you come to wrong conclusions, or you ask the wrong questions. Neurologist Patrick House asks: *"How is it possible to even expect the word 'consciousness' to contain in itself the collapsed variation of billions of years of evolutionary differences? The concept, like the brain that contains it, has evolved."* The concept has definitely evolved, and one of the aspects of evolution is that consciousness is not contained in the brain. High consciousness has evolved within the cultural environment.

Until a more concrete link is achieved between philosophy (the humanities) and science, it will be very difficult to try to know consciousness and time. Speaking of the perspective of scientists, especially mathematicians, Gaston Bachelard believes that: *"The moment one enters the domain of these prophets of the abstract, time is reduced to a simple algebraic variable—the variable par excellence—more apt for the analysis of the possible than the examination of the real. Continuity is, indeed, a scheme of pure possibility for mathematicians, rather than an essential feature of reality."*

As we saw at the beginning, science is born from a historical, Western, Christian (yes,*"Christian"*) view of the universe. Here I repeat something that needs to be made very clear. Science is part of a Western paradigm that comes from creationism. That is to say that, according to these origins, the human being has been created as something separate from the rest of the universe and can study it. For science there is a mind and a reality external to that mind, which is objective reality. Objective reality, I repeat, has been a very useful creation for the advancement of science. That is why, among other things, physics speaks of our everyday reality as something superficial. And it considers it superficial not only because the ordinary human being ignores what happens at the quantum level, but also because it is external to the individual.

Darwin's paradox may lie in having come to the conclusion that the human being is part of the animal kingdom on the basis of a Western and Christian vision. So, in *The Origin of Species*, Darwin applied scientific principles based on an objective reality to study himself, as part of the species.

Today there are some scientists who explain to us that reality does not exist, or that what our perceptions show us has nothing to do with real reality, and the redundancy is meant to be that way. The world is as it is. Evolution has equipped us with the senses we need to survive. Nothing else. Which means that what we perceive is far from reality. You would have to ask a hummingbird, which perceives a million colors, or a dog, which has two hundred and twenty million olfactory cells while humans do not reach five million. Reality would seem to be more than our human perception of it.

But our species has given evolution a second turn. We have outgrown basic, animal, consciousness. How has that happened?

It is clear that if we could perceive reality as it is, perhaps we would not be able to function. On an individual level, we don't even know if other human beings perceive it as we do. We only imagine it. You know what it's like to be in love, to have a headache, to feel what it feels like to see and hear Maria Callas singing *"Casta Diva"*, or to see the moon when it looks bigger or when it turns red. We assume that others have experiences similar to ours. We know there are autistic, or colourblind people. What is obvious is that they are not the norm. Perhaps their perceptions are correct to some extent, or closer to reality than ours. We don't know. All we know is that most people perceive something similar to what we perceive. We may be wrong, but that is the only basis of our communication and it is the best we can hope for. Perhaps there are differences between how we perceive the passage of time and the way others perceive it. Someday someone may be able to define the perception of the burgundy colour and then we will know why Homer defined the Mediterranean as *"that wine-coloured sea"*.

In that sense, time—which is not perceived through the senses—has immeasurable richness. Yes, it is true, it is measured, but with arbitrary, artificial and external means of measuring it, as we have already seen, which are common to make our coexistence possible. However, by wanting to understand its nature through these measurements, all we achieve is to degrade and minimise its essence.

Time is not an illusion. Maybe we'll see the sunrise. And maybe after that, we will have breakfast. And then we'll have lunch. In other words, time passes. And we intuit it, even if it is invisible. Always. But it has an ephemeral quality. We live only the present, the now. That now is extremely complicated and inescapable for human beings, and it is steeped in the past. The now is the axis of our temporal consciousness. As we said, it feeds on the past. How does that happen? Also, is there a precise line that separates past from future? Is our present a thin line, or is it a gelatinous, gooey, sticky reality that is very hard to define? How can music exist in a precise, mathematical, present?

And, as we saw, without the past, we are not. The present also contains instances of the future. Our expectations. Something is going to happen next. We know it, even if what happens is our own death. In the future, some objects that we now know as something relatively stable (the chair, which is not going to disappear) will remain, and processes that have already begun will continue. Or at least, we hope so—it would be strange, and unlikely, if this desk is not here tomorrow, though there's a chance that it will. What we don't know are the events to come. We don't know the episodic. Does that have to do with time?

We know that the mind communicates with the rest of the body through hormones and neurotransmitters. We also know that the mind not only resides in the brain but in the entire nervous system. The connection between nervous system, tendons and muscles is something complex and not fully investigated. But we obviously move. And we act on thoughts.

We said that we do not know the episodic future. But we can imagine possible events. When we fear that the house will burn down, we build ourselves a picture of the burning house in our mind. The imagination lies in that. I dare say that the difference that creative people have is that they can imagine something new and see themselves building it. They take imagination a step further. They make the future.

We could not establish a coherent contact with other people without past, memory and identity.

Scientists, like musicians, or poets, seem to have different sensations of time. Carlo Rovelli, the Italian physicist, says he can't get the idea of particles and the granularity of time out of his head. Quantum gravity tries to explain the ineffable, to understand the unknowable. I get it. But when we analyse a continuum, it seems that we degrade its validity. We cut a rope into a thousand different pieces that are no longer rope and, therefore, it becomes useless. It would appear that scientific knowledge and consciousness have totally different boundaries. We will see that it's not like that at all.

St. Augustine spoke of time as an extension of the soul (high consciousness). Absolutely.

According to St. Augustine, the ability to remember and the ability to have expectations are fundamental elements of high consciousness. One is memory and the other, imagination. Both occur in the context of time. I fully agree with him. According to my personal experience, moreover, without memory-identity there is no future either. Immediately after an amnesiac episode it is impossible to have any expectations, not even remotely imagine the future.

Now psychologists and neurologists speak of subjective time but, according to what we have seen, there doesn't appear to be an objective time either, as there is no objective reality, apart from that created with the sole object of its scientific applications. There is no outside looking in. Or inside looking out. The division between subjective time and objective time is totally artificial and redundant.

Rovelli, as a good scientist, speaks of time as part of the brain, not even of the mind. Understand me well: I think the guy is a genius. But, according to their rules, physicists only observe the measurable aspect, which forces them to start with quantum particles. So—as I said before—for physics, if something can be measured, if it can be quantified, if it can be defined mathematically, if it's an observable quantity, on which other observable quantities depend, then that exists. Consciousness would seem not to exist; time, therefore, does not exist either.

The way I see it, honestly, trying to explain consciousness, time, identity, from an objective reality based on physical particles, is like wanting to describe "Starry Night", Van Gogh's painting, talking about the components of blue pigment. Poor Vincent was crazy and maybe that's why he shot himself. His madness would have been related to chemical reactions or nerve impulses within his brain, and the tortured mind he had. But that same mind, which observed the night and imagined the picture, and the hand that painted it, had no simple relationship with either axons or synapses. Nothing to do with quantum particles. They had to do with an artistic expertise, creativity and sensibility that were part of his personality and genius. From my perspective, the possibility that that—complicated as it is—could be explained by talking about neuronal synapses is, to put it in scientific terms, less than 10^{-44}. And, if it could be done, what validity would it have?

As we have seen, if there is a more precise relationship than that of consciousness-time, it is that of memory-identity. Without memory, we lose identity. Without identity you could not even read what I write.

Returning to time and consciousness. Of course, we have to collaborate. Multidisciplinary (rather than interdisciplinary) research is almost an obligation, but perhaps science and the humanities have to define and differentiate their goals again. Perhaps we have to understand that the validity of the one does not diminish the validity of the others at all. Perhaps we have to return to a Renaissance in which everyone can do everything. Perhaps the absence of a school of innovation defines our current situation. For centuries, specialisation has brought us to this point where ideas are only backed by their provenance. The legal idea of a doctor is as unacceptable as the scientific idea of a lawyer, without going through countless sieves. Brunelleschi was a goldsmith who built the Duomo of Florence. Michelangelo was a sculptor and designed St Peter's. Leonardo was a painter, but he dissected corpses and came to great scientific conclusions.

My first translation, commissioned by the Australian National University in 1970, was an essay on the rods in the compound eyes of caterpillars. As I translated the document—I recall—I thought someone had spent most of their life engrossed in those rods, studying them rather pathologically. The only way our research can advance topics such as awareness and time is through multidisciplinary research, accepting that there may be people who have more than one interest, and creating institutions that promote that individual multidisciplinary interest.

Without renouncing Aristotle, Newton, Einstein, or the possibility of quantum gravity, I believe we have to assume our right to enjoy the song of a nightingale. But, in addition, the search for consciousness and time cannot remain exclusively in the hands of quantum physicists, philosophers and neuroscientists. We all deserve an answer and we all need to work together to find a solution.

Without denying the existence of the Higgs boson, we would have to assume our right to doubt both the Big Bang (as the beginning of all that exists) and God the creator; to own our tears, emotions and to the green of our lawn; understand our finitude and intuit the nature of our very small consciousness and the very short time we have to live on this planet; assume the complexity of how we perceive this world around us and the time that appears to pass within us.

In the next chapter I provide ideas, as a complement to this Conclusion. They are not the ideas of a scientist, or of a philosopher, but I believe they can be of use to both disciplines.

AFTERWORD

A HISTORY-BASED PERSPECTIVE (AND ROADMAP) TOWARDS A NEW APPROACH TO THE "HARD PROBLEM" OF HIGH CONSCIOUSNESS

HYPOTHESES

* Human consciousness consists of two discrete layers:

1) basic animal consciousness

2) high human consciousness.

* High consciousness is only acquired through parental and collective upbringing. It is culturally and individually transmitted.

* The workings of high consciousness cannot be arrived at through the study of basic consciousness.

* Imagination, creativity, language, adventurousness, are exclusively human traits acquired through high consciousness.

* Time is a human device that exists only within high consciousness, through imagination (expectation) and memory (which involves identity and collective perception).

* Without high consciousness there is only present and change.

* Time should not be used as a scientific variable. It is a tool, like a number.

* Science should perhaps consider using 'change' rather than 'time'.

In the 16th Century, Judah Loew, the Rabbi of Prague, used mud from the river and combined different vowels and consonants until he could pronounce the only, the true, name of God. In doing so, he achieved the cabbalistic creation of a living being, the Golem. The creature was meant to be a super human who would save the Jews. It did not happen. The Golem could barely sweep the floor of the synagogue. His brutish, human-created, soul barely allowed him to function. He could not talk.

Three centuries later, in 1816, Mary Shelley took part in an agreement, a competition, with Percy Shelley and Lord Byron to write the ultimate horror story. Mary ended up writing a famous novel about another monster, this time, one created by a gentleman, Dr Victor Frankenstein, who is a mad scientist. Dr Frankenstein is bent on creating an artificial man. The gentleman assembles the monster using body parts. And this creature is another failure.

During part of the twentieth and twenty-first century, computer scientists have worked, and keep on working, tirelessly towards the development of AI, the artificial intelligence based on algorithms that would solve many of the problems of humanity. AI, in its many versions, is a very powerful intelligence, with an amazing, literally superhuman memory. It can beat any human being at chess or go, both games that require a high degree of memory and strategy. AI has helped solve many individual problems that require complicated calculations and formulae. But AI is code-based. It works on the basis of language. AI is a creation of humans that uses a shallow basis of operation. As such, it will never replace the human brain.

The three examples above have similarities. The most important one is that the creations lack the equivalent of human high consciousness or, in the case of AI, any consciousness at all. They may understand some language input, but they do not actually perceive, they do not feel, they do not think the way a human being perceives, feels, and thinks. AI does not even perceive or feel like a living being.

Blade Runner, the movie by Ridley Scott, provides a totally different approach to the subject of humanoids. In this case, the creatures, called "replicants", have a consciousness akin to human *psyche*. They also have an expiry date. They are finite, like us; and have to die.

In a famous monologue, Roy Batty, a rogue replicant who's about to die, tells Deckard, the main character, whose life he has just saved:

"I've seen things you people wouldn't believe... Attack ships on fire off the shoulder of Orion... I watched C-beams glitter in the dark near the Tannhäuser Gate. All those moments will be lost in time, like tears in rain... Time to die".

Great story. It's impossible not to be moved by the scene. The replicant ponders his own finitude and the irretrievable loss of his experiences, and the dove flies away into nothingness.

Human consciousness has evolved strangely, maybe because every individual begins from scratch. Every child learns everything subjectively. Every child is a new hard disk drive. As Borges noticed in *The Witness*, and as Roy Batty tells us, experience is forever lost every time someone dies. One good thing is that humans have notation systems of all sorts. Thankfully, some thoughts survive the individual.

How do we arrive at consciousness? Well, that is a problem that has not been solved so far. I propose that the historical development of the concept of consciousness in the West provides some insights that may help towards its resolution.

First, we would need to accept that there are two types of consciousness, a basic one and a high one, and that they are discrete, evolved from different needs of our species, and evolved during different periods. Studying one type of consciousness will not help solve the problems associated with the other, as they are two different phenomena, discrete and overlapping, not a continuum. This is a matter that complicates the "hard problem of consciousness" as formulated by Chalmers, but that provides a clear path towards its solution, I believe. Of course, consciousness cannot be explained by reducing it to its physical constituents. Qualia are essential, indispensable. All living beings must have them to survive.

The acquisition of high consciousness took some aleatory turns. We humans arrived at it through our relative physical weakness and our need for a lengthy upbringing. We are an altricial species: we are born extremely immature and require maternal and collective care for many years.

The development of this new human consciousness was refined and exponentially accentuated as groups of humans became larger and larger. And it continues to develop and become more sophisticated. The more successful and the more gregarious we became, the more our consciousness grew. In the process, human consciousness acquired elaborate thought, memory, identity and time.

Chalmers' "hard problem" is a dualistic, Cartesian one. A human being can question himself or herself as to why they perceive what they perceive. No other animal can. The "hard problem", I submit, resides in the area of metacognition which, itself, resides in a higher, discrete, layer of consciousness.

Human beings have a basic consciousness because that is the way all living beings relate to their environment. Survival necessitates qualia. That is why humans perceive, feel, and have instincts, just like all other sentient animals.

There are other gregarious species that have some degree of self-awareness and maybe some identity, but they have not developed as successfully as humans have.

Basic consciousness is a primitive form of consciousness, geared towards survival. We have needs. As the animals we are, we need fear, to be applied in "fight or flight" situations; desire, for reproduction; we need to recognise other animals, and danger; we need food to eat, and so on. We have olfactory, auditive, gustatory, visual and tactile perceptions that help us survive. We are born with the ability to perceive. It is a basic ability that develops to some extent like in all other animals. That can be studied through physical analysis and observing neuron oscillations in different parts of the brain. The elements and workings of basic consciousness in humans are beyond the scope of this book.

In humans, however, what happens during childhood is the acquisition of high consciousness through parental and collective upbringing. When attempting to solve the problem of high consciousness, the many studies of the acquisition of consciousness in children appear not to have been taken into account.

In their books, *The evolution of the sensitive soul: learning and the origins of consciousness and Picturing the mind,* Simona Ginsburg y Eva Jablonka, Israeli researchers, propose a precise link between evolutionism and consciousness. Ginsburg and Jablonka study high consciousness in Tel Aviv University. Humans have an unlimited capacity to learn from experience, they both say. Epigenetic variations from one generation to the next do occur as well. Their approach is based on the historic development of consciousness in less developed species. They treat consciousness as a continuum.

Patrick House (*Nineteen ways of looking at consciousness*) ponders the transmission of consciousness: *"If I were asked to create, from scratch and under duress, a universal mechanism for passing consciousness from parent to child, I would probably come up with something a bit like grafting a plant."* ... *"Instead, something much more remarkable happens in nature. An entirely new creature can grow into a fully conscious version of itself, and the entire process occurs, as if by fiat..."*. If we are discussing high consciousness, that is more or less what happens in the case of our species. The child is born before it has developed high consciousness, although with the attributes to achieve it. The child is born extremely immature. A human individual takes several years to develop to the point when it acquires reasoning.

Like other altricial species, humans contribute to development by teaching the child during the period of their upbringing. Human parents and family, clan or community, teach the child words, semantics, syntax, and aid them in the development of their thought processes. Within the community environment, children also develop self-awareness (like in other gregarious species); but also, through memory, they acquire a sense of time that grows into identity, mostly years after the child is given a name.

Truth is, high consciousness is being constructed everywhere on the planet—collectively and exponentially—as I am writing this.

At a certain moment during our childhood, we begin to remember. Our memories—and the development of *a* memory— provide us with a "time" reference that is unique to our species. Thus, high consciousness appears to be a *sine qua non* for the existence of time. I submit that there is no time as such without high consciousness. Do dogs have an idea of human time? No, they do not. They exist from instant to instant and have a basic degree of memory as may be necessary for their everyday lives.

Recently, quantum physicists have been discussing the concept of time as a gooey, sticky, concept. By that, they mean that the present cannot be cut with a knife. Time rolls between past, present and future, but there is no clear boundary. What we understand by present is not an exact millionth of a second, or a very precise instant. Our short-term memory allows us to appreciate music and to remember why we came to the kitchen. All other sentient animals have it. All living beings can tell when abrupt change is happening.

If we discuss the principle of "time" as a human device, something similar appears to apply to other concepts that we take for granted, like "redness", or "reality". A tiger perceives something similar to our red, but has no concept of redness. It has no concept of reality either. It just is, and it just perceives.

Anil Seth (*Being you*) explains reality: *"You could even say that we're all hallucinating at the same time. It's just that when we agree about our hallucinations, that's what we call reality."* Indeed, the objectiveness of "reality", as the objectiveness of "redness" is only provided by the common agreement of the collective. These ideals are only human devices. Anthropocentrism is difficult to avoid. Objective reality is a useful device, but it is not "real" reality—redundance intended.

Why do I assert that time is just a human device? Above, we have assumed that other species possess a limited, short-term, memory. They have a degree of past. But no other animal species can devise a strategy to act, for instance. Only humans can. Wolves can hunt in packs. They use tactics. What humans have, that no other animal has, is a long-term future and a long-term past. In order to have long-term expectations we require imagination; (*ut imaginari*, in Latin, means to be able to create an image in your mind, to visualise a possibility). As we have also said above, our long-term memories allow us an identity.

High consciousness, then, is a more sophisticated type of consciousness that is unique to our species. It has evolved in us because we need it for our interactions with other human beings. It allows us to use language, to communicate, to have language-based thought, to use symbols and codes. It opens up all types of possibilities for co-operation and development that other species cannot reach to the same degree, not even approximately. It also leads to the introduction of time. We developed high consciousness—as I said—through mere luck; mostly because we are born immature, plausibly, because of that we can grow into highly intelligent, gregarious animals; in the words of Aristotle: "*zoon politikon*".

High consciousness interacts with basic consciousness percepts when we use language, for instance. In order to communicate with other individuals, we need to use our auditory and visual abilities. We hear what they say and read what they write. We produce sounds and use codes, like letters and numbers, to communicate with them. And vice versa.

This brings me to how the concept of consciousness developed historically in the West. I have mentioned this above, in the chapter on Greek thought and the Bible. You will have to bear with me if I introduce religion in this equation, but Western understanding evolved through it. I am not a religious person, but there is no escaping from it. Christianity has had an undeniable influence on Western culture. I will try not to make it too lengthy. It does explain many things about the concepts of basic and high consciousness and will help us recognise the difference more clearly.

In ancient Palestine, traditions were transmitted orally until the nation became literate. By the 8th century BC—when Hebrew literacy was introduced—King Hezekiah ordered the compilation of all Jewish myths and legends into the *Tanakh*, a book that would become written tradition, a history, a legal code, the basis of the Jewish religion (Scripture), and the written knowledge on which the two other monotheistic religions would be based: Christianity and Islam. That book would become the most important book in history: the Bible.

There is a Hebrew word: *nephesh* (breath of life), which appears in the *Tanakh* (the Old Testament of the Christian Bible) many times. The early Hebrews thought mostly in terms of "breath of life", rather than basic consciousness. It is plausible that *nephesh* could be equated to the low consciousness of any animal, including a hominin, whereas we could say that *psyche* (a Greek term meaning "soul") applies solely to human consciousness. The former appears in the Old Testament, and the latter, exclusively, in the New Testament. This is the result of a development that led to the Western understanding of the terms. In the *Tanakh*, the ancient Hebrew explanation for the term *psyche* appears to be "knowledge of good and evil". Again, "good" and "evil" are exclusively human concepts.

Plato proposed that our soul (*psyche*) was immortal. As I said, the word *psyche* is used throughout the New Testament to mean "soul". We shall see why.

Nowadays we use the word as "mind" or "consciousness", that is how we know that the concept of consciousness is derived from soul (the same term was used: *psyche*). This was a Greek notion, not explored in Hebrew culture when the Bible originated. It was introduced into Christianity by Saul of Tarsus (aka St Paul, a Hellenistic Jew).

Let's use the historical terms instead of "basic consciousness" and "high consciousness".

St Paul, who was literate, educated and multilingual (he could speak and write in Hebrew, Aramaic, Latin and Greek), and was most probably conversant with some Greek philosophy— especially Plato—interpreted the notion of *psyche*, in Christianity, as an "individual immortal soul".

If *psyche* is considered a collective property of the species, then we can understand why Plato said that it was immortal: we communicate our ideas in space, and in time, from one generation to the other. Our knowledge is shared and, as time passes, it grows.

The Christian innovation was that humans, created by God as separate from other animals, could go to Heaven and be with God. Human individuals became special. Apart from the individual soul, the Christian interpretation of the collective *psyche* of the species became what the Church called the Holy Spirit, part of the Godhead, i.e., the Trinity.

Interpreting Scripture allegorically, the many phenomena that have led to Western development may be clarified. If we apply this method, ancient Hebrews appear to have been proto-evolutionists, that is, I believe they understood evolution. This helps understand *psyche* in Western culture. In the Book of Genesis there are two creations of man: in the first one, God creates man from dust and gives him the breath of life (*nephesh*). In the second one, in the Garden of Eden, human beings acquire Knowledge of Good and Evil (*psyche*).

Rather than a continuum where *nephesh* becomes *psyche*, the phenomenon in the Bible is a marriage of Jewish tradition with Greek philosophy, officiated by St Paul, which leads to the concept of Western consciousness. Two discrete concepts are added one on top of the other and have overlapped ever since. The two-tiered idea of consciousness.

The Garden of Eden allegory is the passage from animal to human, from *Homo Erectus* to *Homo Sapiens*. Two hominins, a male and a female, who were probably part of an original family group or clan (there were many hominins, as we know, wandering the savannas of Africa), become isolated and start speaking to each other. Together with the acquisition of language, they become self-aware and thus modest (they cover themselves, as only humans do). With language, they also acquire *psyche,* and with it, identity, memory, imagination and time. St Paul proposed that Adam and Eve were originally "immortal", i.e., that, as animals they had no concept of time.

I believe the scientific community now acknowledges that the origin of humanity was monogenetic, as proposed in the myth which—we have to bear in mind—was created many centuries before Christ. But the myth does contain a progression and it is a much more sophisticated explanation of high consciousness than previously understood (Pintos-López, R - *The Myth of Adam and Eve and the endurance of Christianity in the West*, q.v.).

We said that, with *psyche*, humans acquired time, which they could not discern as animals, and as such, they could not foresee their finitude (God warned them: "... *you will surely die*"). But enough of religious history.

Aristotle described time by saying that "[it] exists when a soul (*psyche*) can distinguish one moment from another". He also said that "time is the measure of change". As discussed before, time appears to be only the perception of change through *psyche*, so the former definition would be the correct one. Time appears to be just a human, non-physical, percept.

To summarise from what we have seen: in *nephesh* there is the very basic sensory perception that living beings require, i.e., visual, auditive, olfactory, gustatory and tactile perception. All of them translate into percepts, conscious reactions to real physical events or objects. There is self-awareness, limited perception of change: no time, thus, limited memory, no identity, i.e., there is only sensory perception in the present.

Psyche adds another layer that includes the whole gamut of non-sensory percepts: identity, memory, and time. These were originally acquired through upbringing within the family or clan group. The current system includes family and official education.

Perhaps at this point, if I am allowed some self-referencing, I will tell you some first-hand experience of mine to do with self-awareness, identity, memory, and time.

It all began in the year 2001. That afternoon I was on my own at home in Deakin, Canberra, making a silver pendant for Inés, my wife, who had gone to the USA to complete her postgraduate studies in psychology. While working with the silver, I had to "anneal" the sheet of silver (heat it to make it softer and thus being able to work it more easily). After turning the torch off, I had an almost mystical experience. I perceived an exquisite women's perfume. My explanation at the time was that it probably had something to do with the peculiar smell of the silver when heated up to that extreme. I went to the other side of the house to make myself a cup of coffee, hoping for the strange experience to end. The aroma followed me for a while, even after having made the cup of coffee, in another room, and several metres away from my workshop. Weeks later I discovered that the olfactory hallucination was called "phantosmia". I would much later find out that the episode was the beginning of a pathology of the mind. Phantosmia was the preamble to something larger, equally strange, and not as pleasant.

One morning, years later, I woke up without identity, without past or future. I was living in a weird present. How could that be? Well, it was dawning; half asleep, I looked at the ceiling and, later towards the wall where we have the TV set, and I asked myself: *"Where am I?" "Who am I?"*. Lying in bed, next to me was Inés. I said to myself *"This woman must have something to do with me, because she is sleeping next to me "*. At the time I didn't have the faintest idea of who I was, where I was, that Australia, or Queensland existed, or that I was originally from Argentina. I didn't know my age either. I was in bed, somewhere. That, I knew. Maybe I could tell that I was a human being. I don't remember if my thoughts were in English or in Spanish. I believe they were in English. I didn't have the slightest memory of anything, and I could not begin to think what, if anything, would happen afterwards.

After a short while, I came out of the episode and, as I recovered memory and identity, I started chatting with Inés. We laughed for a few minutes about the thought: Can you imagine not knowing who I was or where I was? What had happened, we said, was that I was not quite awake, that was all.

A few weeks passed. I was in Sorrento, with one of my sons and daughter-in-law in a holiday house, on the coast of Victoria. Again, I woke up in the middle of the night without recognising the place. The room had a nautical decor. On the wall there was an oar. I did not understand very well what it was, or what it was for. I got up and started walking along a strange corridor without knowing where it would lead to, who I could call, and without understanding if there was another person who would answer back. I'm not easily scared, but the experience was terrifying. When somebody asks me how it feels not to know who you are, having turned up in a strange present, in a strange place, my answer is: *"It's like—I imagine—an astronaut would feel, untethered to the spaceship and floating, lost in space; but different to some extent, as the astronaut would know who she was and what she was doing."* Perhaps it would be better to explain it as somehow disappearing from where one is and appearing, by some kind of magic, in a restaurant in Berlín, or in Istambul's main market, for instance. But without knowing who you are or why you are there. Something totally unheard of and absurd at the same time.

In my case, the "absences" kept on occurring. There were four or five more seizures. One of them happened after being driving for a while and having parked the car at a shopping centre that was totally unknown to me, and that —of course—I knew perfectly well. My GP told me that it was probably some kind of epileptic amnesia. I went to see a psychologist (who could not help me), underwent some tests and went to see a neurologist several times. It was *TEA* (*Transient Epileptic Amnesia*). With medication, the condition disappeared and remains luckily under control. The neurologist explained that sometimes phantosmia is a prelude to future epileptic seizures.

As it turns out, now I know firsthand, how you feel when, after losing consciousness, you wake up without identity or a sense of time. After waking up from one of these episodes, there is a space-time in which, without *psyche*, you are not there. You have self-awareness but no identity.

That is similar to what happened to Clive Wearing. He was born in England in 1938 and was a very successful musicologist, pianist and conductor. I say "was" because, although he is still alive, none of those things are anymore. He was married and had children. In the eighties he had a mild herpes viral infection. The infection passed into the bloodstream and from there to the nervous system, with such bad luck that, when it reached the brain, it affected the hippocampus, which is the area where short-term memories are transmitted to long-term memories. That resulted in a unique case of simultaneous, anterograde and retrograde amnesias, meaning that, Wearing, at present, remembers very little of his past and also cannot form new memories. The memory he has left only serves to remember between twelve and thirty seconds. He spends every day of his life waking up. One can talk to Wearing, but after thirty seconds at most, he thinks he has woken up from a comatose state. He converses, acts normally, and after thirty seconds everything starts again. He knows he has children, but he doesn't remember their names. He knows how to groom himself, and dress, and communicates in perfect English. He knows how to play the piano impeccably, and remembers being totally in love with his second wife, Deborah. That's all. Of course, he cannot live a normal life and has been hospitalized for more than forty years. He keeps a diary in which he writes similar things every day: *"I feel very good. I woke up for the first time."* In an interview with three journalists in 2005, he was asked how he felt, if he was sad or happy.

He replied, *"You are the first human beings I see, the three of you. Two men and one woman. The first people I see since I got sick. There is no difference between day and night. I don't think about anything. I have no dreams. Day and night, always the same: everything is blank. Exactly like being dead." "Is it very difficult to live like this?"* —they asked. *"Nope. It's exactly the same as being dead, which isn't difficult, right? Being dead is easy. You don't do anything. One cannot do anything when one is dead. It's all the same. Exactly."* They asked him: *"Do you miss your previous life?" "Yes, but I'm never conscious enough to think about it. I have never been bored or grieving. I have never felt any way. It is the same as death. I don't even have dreams. Day and night, everything is the same." "When you miss your previous life, you say, 'Yes, I miss my previous life,' what do you miss?" "That I was a musician. And that I was in love."* The thing about Wearing is that he doesn't possess a sense of time. Literally, time does not exist for him. When he says it's just like being dead, that's exactly it. When he returns from his amnesiac state it is as if an eternity has passed. It is like being in a time that exists outside the type of life we know: like being dead or not having been born.

It is like sleeping very profoundly, without dreams. Afterwards, one wakes up with amnesia. You are there, but you are not yourself. There is some consciousness, and some self-awareness, but no identity (without memory, there is no time, and without time, there is no identity, past or future). What is terrifying is that, being basically conscious there are no expectations either. Because of that, there is no idea of finitude. One is not a human being.

In order to clarify this: at the time of waking up from an amnesic episode, one has a semblance of consciousness. The waking up itself becomes that germ of consciousness. And one is obviously self-aware. Memory returns slowly, and with it, one's identity. With one's memories, one regains, rather than self-awareness, selfhood, and the quality of being human, if that makes sense. With this, I have to conclude that you are the sum of your memories. And also —perhaps it could be added—the sum of your dreams.

Anil Seth's "beast machine theory" fails to make this difference. He tells us: *"... consciousness is more closely connected with being alive than with being intelligent. Naturally, this applies as much to other animals as it does to us humans. On this view, consciousness may be more widespread that it would seem, were we to take intelligence as the primary criterion."* But the difference does not appear to lie between life and intelligence. Seth's approach is based on a continuum of consciousness, whereas what was necessary to reach *psyche* was a quantum leap from one layer to another. The Garden of Eden experience. The allegory that attempts to explain the beginning of humans in society.

What is extraordinary is not that identity involves memory. That is well known and understood. With *nephesh* it is possible to have self-awareness. But to have a human identity you do not need just *nephesh*, you need *psyche*. With your identity you exist in time. That includes your past and also your future. The price we pay for identity is finitude.

We were saying that *nephesh* and *psyche* are the equivalent to two distinct layers of perception, a sensory layer and a non-sensory one. The first layer is shared by all animals. The second layer of perception is acquired individually, and is exclusively human.

Albert Einstein's discovery of relativity based on the unity of space-time was a stroke of genius in that he counter-intuitively combined the sensory and non-sensory components of space-time. He never accepted another counter-intuitive discovery, though: Einstein rejected the Uncertainty Principle in quantum mechanics until he died.

I submit that a study of encephalic activity related to the first layer of perception will only elucidate questions such as how physical perception leads to basic decision-making in individuals, but will fail to lead to a solution of the "hard problem", i.e., "why do I feel the way I feel about this?". Likewise, a study of the development of consciousness based on evolution from more primitive to more advanced animals will only reach the *nephesh* "ceiling". Without questioning their scientific validity, I believe both types of study are limited by the discreteness of layering in human consciousness.

In all probability, the way in which the "hard problem" will be solved—I suggest—is through the study of consciousness from a current microcosmic perspective, i.e., the study of the acquisition of consciousness in children. The approach required cannot be either a physical one, or an evolutionary one, as they would both involve only the study of *nephesh*, not of *psyche*. The phenomena are overlapping but not identical. Labyrinths sometimes lead to dead ends.

Psyche is currently and continuously evolving in children. A detailed study of the process would open a window, a new approach to Chalmers' "hard problem".

ACKNOWLEDGMENTS

ACKNOWLEDGEMENTS

I owe a huge debt of gratitude to Inés, my wife, who read the whole manuscript several times as it evolved.

Friends in Australia read parts, as happened in New Jersey, Chicago and San Francisco.

I want to thank the words of encouragement I received from Tarah and Rod Haedo, who recommended the book by Carlo Rovelli.

Professor Ezequiel Morsella found time in his busy schedule to read part of the Introduction. I thank him.

Dr David Barbour reviewed a shorter version of the Afterword and provided some very useful insights. I thank him.

My brother Patocho also read parts of the book.

Mark Lush, again, found time to read the afterword, which is really a summary of the hypotheses included in the book.

www.ingramcontent.com/pod-product-compliance
Lightning Source LLC
Chambersburg PA
CBHW051439290426
44109CB00016B/1617